PRACTICAL MANUAL
IN
BOTANY
VOLUME I
FOR
DEGREE STUDENTS

PRACTICAL MANUAL IN BOTANY

By
Poonam Sethi
Asst. Professor, Guru Nanak
College, Chennai, India

Foreword By
D. Kalaiwani
Head of Botany Department,
Presidency College, Chennai, India

GenNext Publication

ISBN : 978-93-8022-255-4

Price : ₹ 350

First Published, 2015

Published by

genNext Publication
23, Main Ansari Road, Darya Ganj
New Delhi-110002
Ph. : 23242060, 23261060
Fax : 011-23285914
E-mail: gennext@hotmail.com
Website : www.gyanbooks.com

Laser Typesetting : Gyan Prints (AB-I), Delhi
Printed at : Isha Print.

Cataloging in Publication Data--DK
Courtesy: D.K. Agencies (P) Ltd. <docinfo@dkagencies.com>

Sethi, Poonam, author.
Practical manual in botany / by Poonam Sethi ; foreword by D. Kalaiwani.
volume 1 cm
ISBN 9789380222554

1. Botany—Textbooks. I. Title.

DDC 580 23

Dedicated to

MY SATHGURU

AND

SAI BABA

Contents

Contents

Preface

The present book is an outcome of encouragement shown by Professors and students of BOTANY mainly Under graduates. This book caters to the laboratory needs of I Bsc. and II Bsc. students of nearly all the universities in the country

The book is written in a simple language and its approach has been student oriented with photographs and illustrations.

I shall welcome constructive suggestions from its readers.

Poonam Sethi

Dept. of BOTANY

GURU NANAK COLLEGE, CHENNAI.

Foreword

I have a pleasure in writing a foreword for this book entitled PRACTICAL MANUAL IN BOTANY', Vol. I. My friend and student Asst Prof. POONAM SETHI, author of the book has to her credit years of teaching and research experience.

This volume is written in lucid style and would meet the need of students preparing for the Bsc. degree of Indian Universities

I am confident that this volume will be received well by both student and teachers and will prove popular.

My hearty congratulations to the author on has achievement.Desination.

D. Kalaiwani

Head of the Department

Presidency College, Chennai

About The Author

Dr. POONAM SETHI, professor in botany (PBPBT) GURU NANAK COLLEGE, CHENNAI, passed Msc. (1999) in first division with university Rank (university of Madras) and was also awarded the Best out standing student of the year. Pulsed ph. D in PHYCOLOGY and has to her credit a compound from seaweeds against Human Coxsackie virus. She has published over 15 research papers in reputed Indian and foreign Research journals.

Has experience of teaching Botany to the under Graduate student and organised a workshop on screening and control of Viral diseases. Endowed with passion for scientific writings has written several books for various competitive exams.

This Manual has been written in a student friendly manner and presented in a simple and lucid style well supported with self explanatory illustrations, immensely useful for Botany students.

PRACTICAL MANUAL
IN
BOTANY
Volume I
For
DEGREE STUDENTS

(ALGAE, FUNGI, LICHENS, BACTERIA, VIRUSES, BRYOPHYTES, PTERIDOPHYTES AND GYMNOSPERMS)

Introduction

Botany is the study of plants, and they provide an essential foundation for life on earth, the food we eat and the beauty of the natural environment botany as a result is considered to be an extremely important science! Botany is no longer confined to the study of how and why plants survive in the way they do. With the advancement of technology, the study of botany means the study of how plants are used in biotechnology, the significant part that they play in ecosystems and of course, how they can continue to provide food, medicines and even fuel to an ever-expanding global population. An holistic approach to understand the general characters of each genera in the syllabus mainly of lower cryptograms is done with utmost care.

This book is written in simple language and its approach has been student oriented. It is profusely illustrated with photo-micrographs and line diagrams which serves the entire purpose of practical needs and requirements of under-graduate students of all universities.

Lastly, I shall welcome constructive suggestions from readers.

(I) MICROSCOPE

A microscope is a tool to look or see objects that are too small for the naked eye. The science of investigating small objects using such an instrument is called microscopy. Microscopic means invisible to the eye unless aided by a microscope.

(i) Types

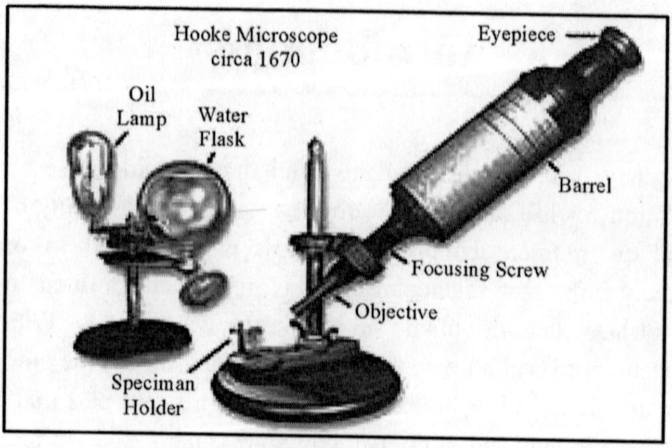

Fig. Ancient Microscopes

The first microscopes were produced around the same time as the telescope. They also had two or more lenses, but with an objective lens of shorter focal length.

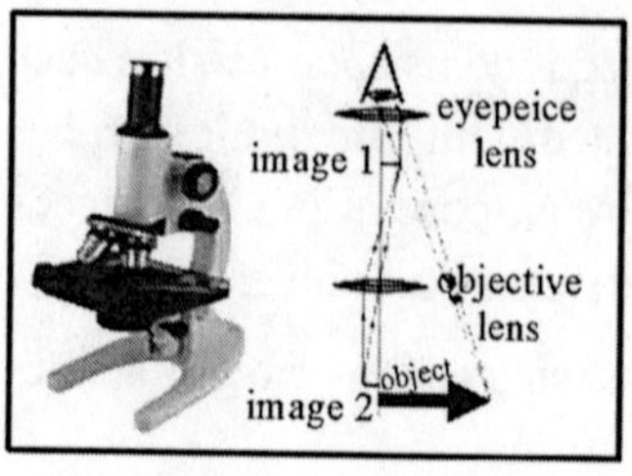

1. Compound Microscope

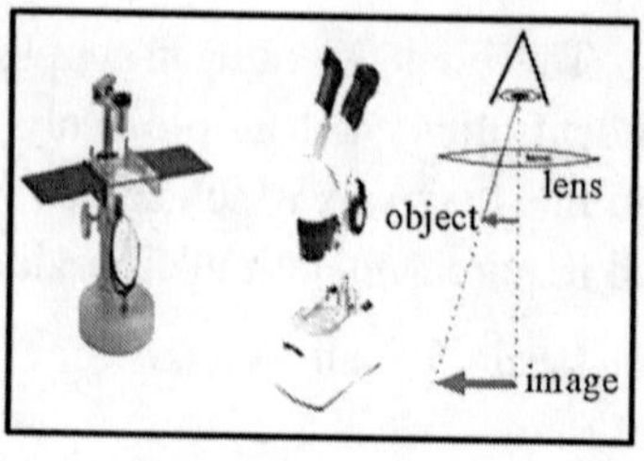

2. Dissection Microscope

The *compound light microscope* is popular among botanists for studying plant cells, and view bacteria and parasites as well as a variety of human/animal cells in biology. It is a useful microscope in forensic labs for identifying drug structures, combines the power of lenses and light to enlarge the subject being viewed.

A *dissection microscope* is a microscope that uses lens to enlarge an object through angular magnification, giving the viewer an erect enlarged virtual image. Simple microscopes are not capable of high magnification.

The transmission electron microscope (TEM), the first type of EM, is capable of producing images one nanometer in size.

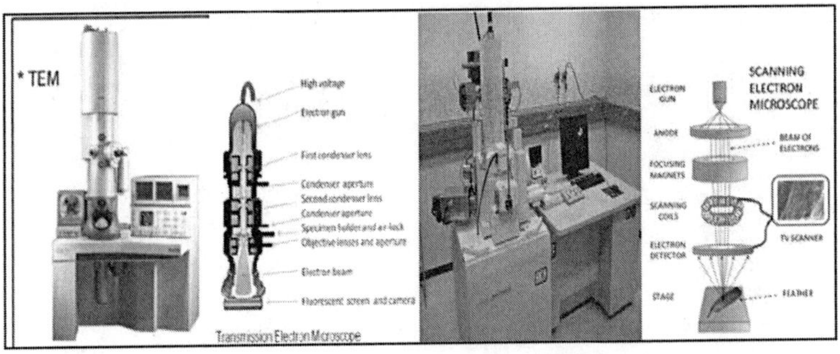

3. Electron Microscope (EM): TEM and SEM

The TEM is a popular choice for nanotechnology as well as semiconductor analysis and production. A second type of electron microscope is the scanning electron microscope(SEM) are approximately ten times less powerful than TEMs, they produce high-resolution, sharp, black and white 3D images.

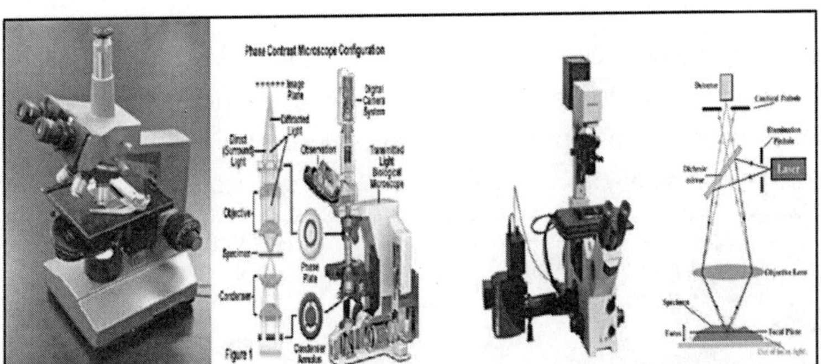

4. Phase Contrast Microscope **5. Confocal Microscope**

Phase contrast microscopy allows the viewing of unstained specimens by using the light phase amplitude differences within microscopic objects. Phase differences, between light passing through a specimen and that passing through surrounding medium, into brightness differences provides a difference in brightness between the object and the background, which the eye can see.

Confocal microscopy is an optical imaging technique used to increase optical resolution and contrast of a micrograph by using point illumination and a spatial pinhole to eliminate out-of-focus light in specimens that are thicker than the focal plane, it enables the reconstruction of three-dimensional structures from the obtained images.

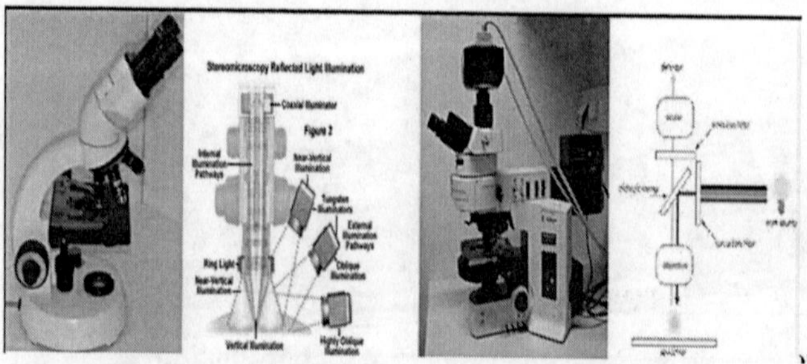

The *Stereo microscope*, also called a dissecting microscope, has two optical paths at slightly different angles allowing the image to be viewed three-dimensionally under the lenses. Stereo microscopes magnify at low power, typically between 10X and 200X, generally below 100x.

A *fluorescence microscope* is an optical microscope with fluorescence and phosphorescence instead of, or in addition to, reflection and absorption to study properties of organic or inorganic substances. Refers to any microscope that uses fluorescence to generate an image, a more simple set up, like an epifluorescence microscope.

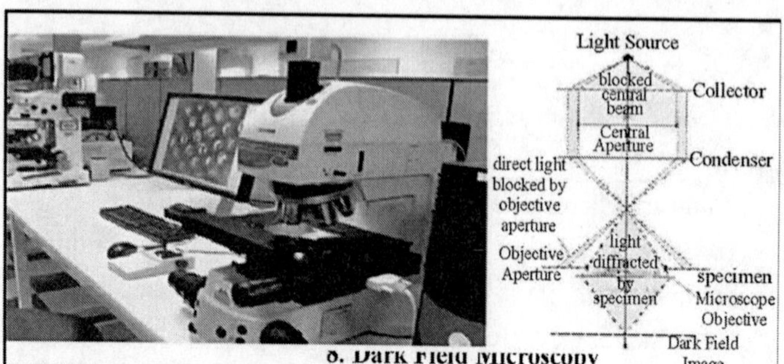

8. Dark Field Microscopy

Dark field microscopy describes in both light and electron microscopy which exclude unscattered beam from the image, the field around the specimen is generally dark.

(ii) Parts of Compound Microscope or Students Microscope

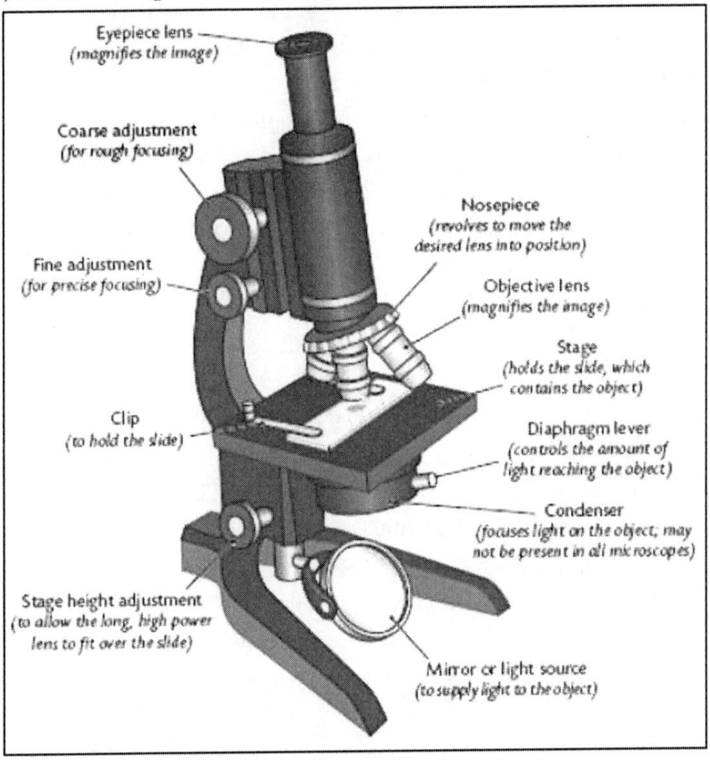

Fig. Parts of a Compound Microscope

The essential parts of a compound microscope are:

Foot and Limb—Bottom most part of the microscope foot from which arises the limb.

Stage—Stage is borne on the limb, it the objects are placed on to be examined, two stage clips are provided.

Mirror—Located below the stage, functions to reflect light and illuminate objects.

Body—It is a brass tube which is movable by the rack and pinion mechanism vertically. This is coarse adjustment.

Fine Adjustment—In use of high powers, fine adjustments are utilised.

Lens system—The eye piece or ocular system is fixed on the upper side. Objectives are screwed at the lower end. Function of objective is to form enlarged image of the object, whereas eye piece magnifies this image. The sum of magnifying powers of both objective and eyepiece constitutes the total magnifying power of the microscope. This can be varied by changing one or the other. Eye piece should be either 3X, 5X, 8X, 10X, or 15X, having magnification of 3, 5, 8, 10, 15 times. Objectives should be low power (2/3' focal length) or high power (1/6'focal length).

(iii) How to Use

1. Inspect the Foot and Limb. The limb holds the optical elements, which include the eyepiece and eyepiece tube, the nose piece, and the objective lenses. Study the arm. The arm connects the head to the base. The microscope is usually carried by the arm.
2. Examine the Base. The base holds up the microscope and provides the stage onto which samples are placed. The base also contains the illuminator and the focusing knobs (both fine and coarse).
   - Position your sample between two pieces of glass to make a slide.
   - Place the slide on the center of the stage over the glass hole.
   - Move the two stage clips over the sides of the slide to secure it into place.
3. Turn the nose piece turret until the shortest lens is above your sample. It should click and feel rigid when it is in place. The shortest objective lens is the least powerful and is the easiest level to begin when magnifying an object.
4. Twist the coarse focus knob (the large one) on the side of the base so the stage moves upward towards the short objective lens. Perform this without looking into the eyepiece. It is important to

make sure the slide does not touch the lens. Stop twisting the coarse knob just before the slide contacts the lens.

5. Look through the eyepiece.
6. Arrange the illuminator until you've arrived at a comfortable level of light. The brighter the illuminator shines, the better you could see your specimen, but set to the maximum brightness if it hurts your eye.
7. Twist the coarse focus knob the opposite way as you had before, so the stage moves away from the lens. Do this slowly until the sample begins to focus.
8. Turn the fine focus knob to perfect the focus in greater detail.
9. Reposition the slide as necessary to situate the area of desire study directly under you gaze.
10. Switch lenses by rotating the nose piece turret to a longer lens. Perform necessary focusing adaptations.

Turn off the power, lower the stage, remove your specimen and cover the equipment with a dust-repellent cover.

(iv) Microscope Magnification

The unit of measurement of a micro organism is μ (1/1000 mm). The ocular micrometer and stage micrometer are fixed below the eye piece and on the stage respectively. Each division of the stage scale is equal to 0.01mm or 10μ. Certain ocular micrometer covers certain divisions of the stage micrometer. For example, x divisions of ocular scale coincide with the y divisions of stage scale.

So, one division of ocular coincides with y/x of stage.

1 division of stage=0.01mm or 10 μ

So y/x=0.01*y/x mm

Since, 1/1000mm=1 μ

So, 0.01*y/x =0.01*y/x *10

Now place the object on the slide and fix the ocular scale. Measure the length, breadth and diameter in terms of the divisions of ocular scale. This number is multiplied by power of the objective as calculated above. It denotes the measurement of the object in μ.This method measures the object, as it is original in size.

The magnified image is calculated as follows:

$$\text{Magnification} = \frac{\text{Magnification of eye piece * Length of tube in microscope}}{\text{Focal lenth of the objective lens}}$$

Tube length = 160mm

(II) STAINS USED IN MICROSCOPY

Staining is an auxiliary technique used in microscopy to enhance contrast in the microscopic image. Stains and dyes are frequently used in biology and medicine to highlight structures in biological tissues for viewing, often with the aid of different microscopes.

(i) Some Common Biological Stains

Acridine Orange: Acridine Orange (AO) is a nucleic acid selective fluorescent cationic dye useful for cell cycle determination. It is cell-permeable, and interacts with DNA and RNA by intercalation or electrostatic attraction. When bound to DNA, it is very similar spectrally to fluorescein.

Bismarck Brown: Bismarck brown (also Bismarck brown Y or Manchester brown) imparts an yellow colour to acid mucins.

Carmine: Carmine is an intensely red dye used to stain glycogen, while Carmine alum is a nuclear stain. Carmine stains require the use of a mordant, usually aluminum.

Coomassie Blue: Coomassie blue also brilliant blue non-specific stains proteins a strong blue colour. It is often used in gel electrophoresis.

Cresyl Violet: Cresyl violet stains the acidic components of the neuronal cytoplasm a violet colour, specifically nissl bodies. Often, it is used in brain research.

Crystal Violet: Crystal violet, when combined with a suitable mordant, stains cell walls purple. Crystal violet is an important component in Gram staining.

Eosin: It is most often used as a counterstain to haematoxylin, imparting a pink or red colour to cytoplasmic material, cell membranes, and some extracellular structures. It also imparts a strong red colour to red blood cells. Eosin may be used as a counterstain in some variants of gram staining, and in many other protocols.

Ethidium Bromide: Ethidium bromide intercalates and stains DNA, providing a fluorescent red-orange stain. Although, it will not stain healthy cells, it can be used to identify cells that are in the final stages of apoptosis—such cells have much more permeable membranes

Acid Fuchsine: Acid fuchsine may be used to stain collagen, smooth muscle or mitochondria. Acid fuchsine is used as the nuclear and cytoplasmic stain.

Haematoxylin: Haematoxylin is a nuclear stain. Used with a mordant, haematoxylin stains nuclei blue-violet or brown. It is most often used with eosin in H&E (haematoxylin and eosin) staining—one of the most common procedures in histology.

Iodine: Iodine is used in chemistry as an indicator for starch. When starch is mixed with iodine in solution, an intensely dark blue colour develops, representing a starch/iodine complex. Starch is a substance common to most plant cells and so a weak iodine solution will stain starch present in the cells.

Malachite Green: Malachite green also known as diamond green B or victoria green B can be used as a blue-green counterstain to safranin in the Gimenez staining technique for bacteria. It can also be used to directly stain spores.

Methyl Green: Methyl green is commonly used with bright-field microscopes to dye the chromatin of cells so that they are more easily viewed.

Methylene Blue: Methylene blue is used to stain animal cells, such as human cheek cells, to make their nuclei more observable. it is also used to stain the blood film and in cytology.

Neutral Red: Neutral red or toluylene red stains Nissl substance red. It is usually used as a counterstain in combination with other dyes.

Nile Blue: Nile blue or Nile blue A stains nuclei blue. It may be used with living cells.

Nile Red: Nile red also known as Nile blue oxazone is formed by boiling Nile blue with sulfuric acid. This produces a mixture of Nile red and Nile blue. Nile red is a lipophilic staining accumulat in lipid globules inside cells, staining them red. Nile red can be used with living cells. It fluoresces strongly when partitioned into lipids, but practically not at all in aqueous solution.

Osmium Tetraoxide: Osmium tetraoxide is used in optical microscopy to stain lipids. It dissolves in fats, and is reduced by organic materials to elemental osmium, an easily visible black substance.

Safranin: Safranin or Safranin O is a nuclear stain. It produces red nuclei, and is used primarily as a counterstain. Safranin may also be used to give an yellow colour to collagen.

(ii) Staining Technique

It is necessary to stain microorganisms before they can be viewed with the light microscope for microbial cytoplasm It is usually transparent, staining is unnecessary when microorganisms are very large and a drop of the microorganisms can be placed directly on the slide and observed. A preparation such as this is called a wet mount. A wet mount can also be prepared by placing a drop of culture on a coverslip and then inverting it over a hollowed out slide. This procedure is called the hanging drop.

Simple stain techniques can be performed with basic dyes such as crystal violet or methylene blue, positively charged dyes that are attracted to the negatively charged materials of the microbial cytoplasm. Such procedure is the *simple stain procedure.* An alternative is to use a dye such as nigrosin or Congo red, acidic, negatively charged dyes.

They are repelled by the negatively charged cytoplasm and gather around the cells, leaving the cells clear and unstained. This technique is called the *negative stain technique.*

The *differential stain technique* distinguishes two kinds of organisms. An example is the *Gram stain technique.* This differential technique separates bacteria into two groups, Gram-positive bacteria and Gram-negative bacteria. Crystal violet is first applied, followed by mordant iodine, which fixes the stain. Then the slide is washed with alcohol, and the Gram-positive bacteria retain the crystal-violet iodine stain; however, the Gram-negative bacteria lose the stain. The Gram-negative bacteria subsequently stain with the safranin dye, the counter stain is used next.

These bacteria appear red under the lens, while Gram-positive bacteria appear blue or purple, reflecting the crystal violet retained during the washing step.

Tissues which take up stains are called chromatic. Chromosomes were so named because of their ability to absorb a violet stain. Positive affinity for specific stain may be designated by the suffix-*philic.* For example, tissues that stain with an azure stain may be referred to as azurophilic.

This may be used for more generalised staining properties, such as acidophilic for tissues that stain by acidic stains (most notably eosin), basophilic when staining in basic dyes, and amphophilic when staining with either acid or basic dyes. In contrast, Chromophobic tissues do not take up coloured dye readily.

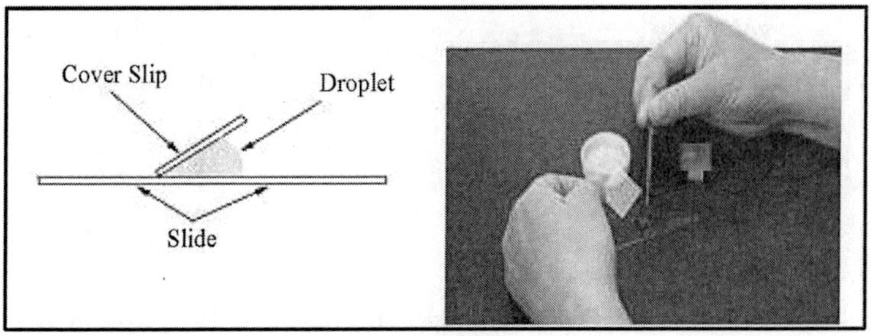

Fig. Staining Technique

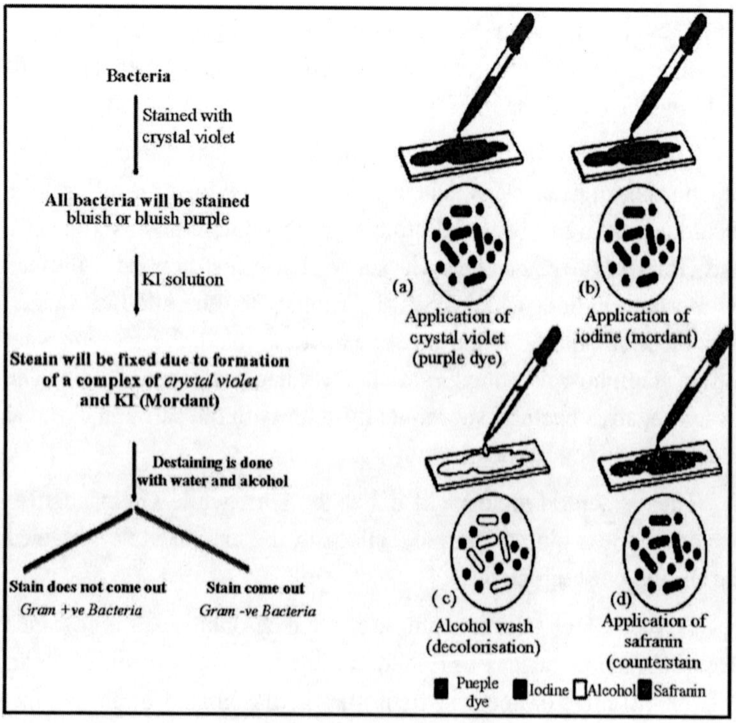

Fig. Gram Staining Technique

(III) TECHNIQUE OF TAKING SECTION

(a) *Materials*

1. Razor Blades
2. Forceps
3. Watch Glass for Staining
4. Clean Slides and Coverslips
5. Toluidine Blue (0.05 per cent aqueous) and other stains as necessary
6. Dropper Bottle containing water
7. Waste Cloth

(b) *Procedure*

1. Sit comfortably with forearms resting on the bench and elbows close to your sides. Hold plant tissue between your thumb and forefinger.

2. Wet razor blade, fingers and tissue with water from the watch glass. Water should drip from your fingers during sectioning.
3. Cut the tissue quickly and smoothly in the plane desired. Now section slowly by drawing the razor blade towards you in a smooth slicing motion; the razor should rest on the tip of your thumb. Use your thumb to control the thickness and evenness of the sections. This takes practice. Concentrate on getting very thin portions of some sections. It is not necessary to obtain complete cross-sections.
4. Transfer sections to a depression containing toluidine blue and stain for 15 seconds. Do not use stain that has evaporated and begun to precipitate.
5. Rinse sections in a watch glass containing water.
6. Mount sections on clean slides in a drop of water. To apply the cover slip, hold at an angle and touch the water drop with one edge. Lower the cover slip slowly to avoid air bubbles. Semi-permanent mounts can be made by fixing tissue in phosphate-buffered glutaraldehyde, mounting in glycerol jelly, and sealing the cover slip with nail polish.

(c) *Types of Section*

(i) T.S. or V.S. or cross-section

A section which cuts the material at right angles is a cross-section. When the vertical axis is cut at right angles, it is a transverse section, but when the section is parallel to earth surface it is a vertical section.

(ii) T.L.S. or Tangential Longitudinal Section

In this section, it cuts the circumference at right angles to the radius that is tangential.

(iii) R.L.S. or M.L.S. or Radial Longitudinal Section

In this, the section passes through radius, but when it passes parallel to diameter, it is called as median longitudinal section.

(IV) PLANT KINGDOM

The plant kingdom is made up of multicellular, photosynthetic eukaryotes that evolved from green algae. There are 10 divisions within the kingdom, which differ with respect to morphology and reproductive anatomy. All plants have chloroplasts and cell walls, and go through an alternation of generations involving diploid sporophyte and haploid gametophyte. However, plants differ in the presence or absence of three important structures: a vascular system, seeds, and flowers/ fruits. These adaptations aided plants in the colonisation of land from aquatic ancestors, and surviving in the changing terrestrial habitats. The description below lists different evolutionary steps in the plant phylogeny.

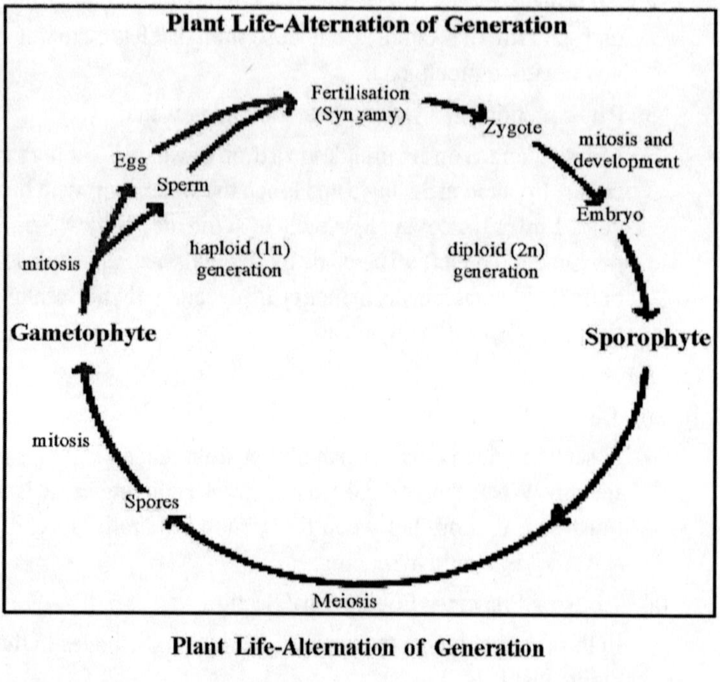

Plant Life-Alternation of Generation

Basically the thallophyta are the plants where the plant body is a thallus, largest division of the plant kingdom with about 61,000 species. Algae, 19000, fungi and bacteria 12,000. Plants are simplest and primitive with similar cells. Plants are not organised into root stem and leaves no cellular differentiation. The thallophytes include algae, fungi, bryophytes, pteridophytes but the gymnosperms, naked seed plants, form a separate group followed by angiosperms, enclosed seed plants.

PART II

Type Study

I ALGAE

(I) Chlorophyceae

(a) Chlamydomonas

Systematic Position

Class—Chlorophyceae

Order—Volvocales

Sub Order—Chlamydomonadineae

Family—Chlamydomonadaceae

Genus—Chlamydomonas

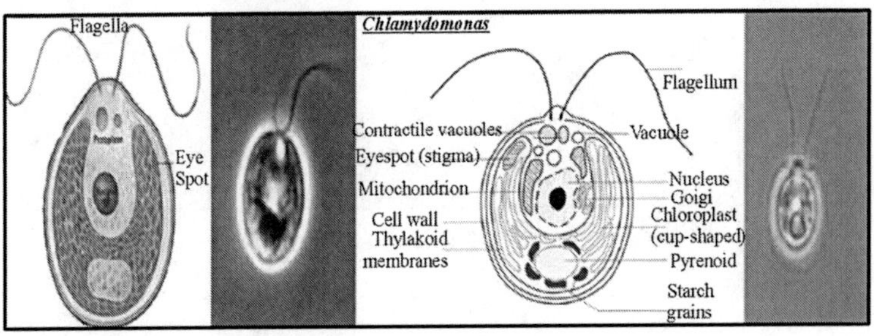

Fig. A Typical Cell of Chlamydomonas

1. It is a unicellular green algae, unicellular flagellates, occurs in ponds, ditches, pools and lakes.
2. It is spherical, ovoid, pyriform or ellipsoidal.
3. Anterior end-pointed, papilla present, cell wall made up of cellulose Flagella originates from flagellar or neuromotor apparatus. flagella arise from basal granules, two contractile vacuoles are present beneath basal granule.
4. The most prominent structure is the chloroplast, horse shoe or cup shaped, pyrenoid embedded in the chloroplast, stigma or eye spot present. It is generally found in habitat rich in ammonium salt. The red eye spots for photosensitivity.

5. Reproduction both sexual and asexual. Chlamydomonas asexual reproduction goes through zoospores, palmella stage, alpanospores and hypnospores; sexual reproduction through isogamy, anisogamy or oogamy.

(b) Volvox

Systematic Position

Class—Chlorophyceae

Order—Volvocales

Family—Volvocaceae

Genus—Volvox

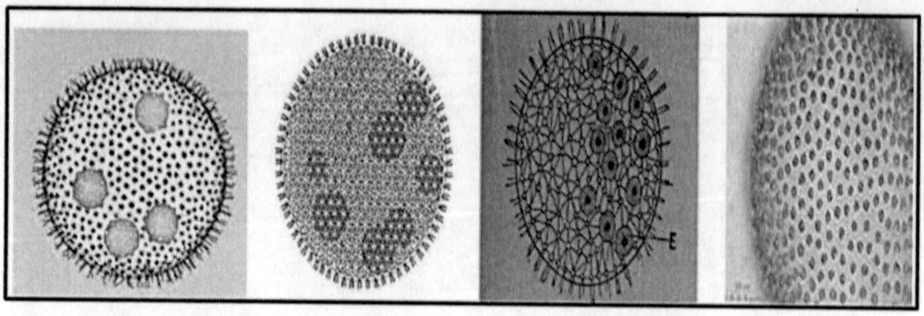

Fig. A colony of Volvox-Coenobium

1. It is a fresh water algae, colonial form, visible as a greenish round body.
2. Colony or coenobium containing an extracellular matrix made of a gelatinous glycoprotein, each cell typically chlamydomonas type interconnected by protoplasmic threads or thin strands of cytoplasm, called protoplasmates.
3. Each cell has a chloroplast, pyrenoid, contractile vacuole, eye spot and a pair of cilia, bicillliate, arising from two basal granule. The whole colony moves by the concerted action of all cilia.
4. Reproduction of two types asexual and asexual, vegetative reproduction absent.
5. Asexual reproduction during favourable conditions and sexual during unfavourable conditions, sexual reproduction oogamous.

In sexual reproduction two types of gametes are produced. It can be monoecious or dioecious. Male colonies release numerous microgametes, or sperm, while in female colonies single cells enlarge to become oogametes, or eggs.

(c) Chlorella

Systematic Position

Class—Chlorophyceae

Order—Chlorococcales

Family—Chlorocoocaceae

Genus—Chlorella

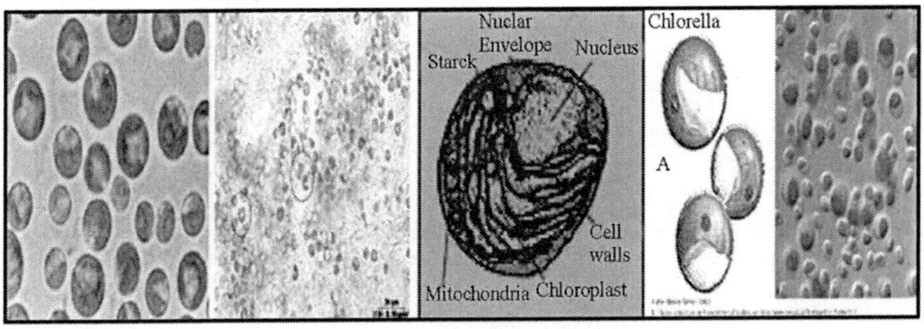

Fig. Chlorella under Microscope

1. Chlorella is a genus of single-cell green algae. It is spherical in shape, about 2 to 10 mn in diameter and is without flagella. The name chlorella is taken from the Greek chloros, meaning green, and the suffix *ella*, meaning small.
2. Chlorella contains the green photosynthetic pigments chlorophyll—(a) and (b) in its cup shaped chloroplast, single pyrenoid present, exceptionally high concentration of chlorophyll.
3. It is an attractive potential food source because it is high in protein and other essential nutrients; when dried, it is about 45 per cent protein, 20 per cent fat, 20 per cent carbohydrate, five per cent fibre, and 10 per cent minerals and vitamins. It is also abundant in calories, fat, and vitamins.
4. Its reproduction is asexual by nonmotile reproductive cells, autospores. Sexual reproduction absent.

5. Mass-production methods are now being used to cultivate it in large artificial circular ponds, *Chlorella* multiplies rapidly and is rich in proteins and B-complex vitamins; it has also been studied as a potential food product for humans both on Earth and in outer space.

(d) Scenedesmus

Systematic Position

Class—Chlorophyceae

Order—Chlorococcales

Family—Scenedesmaceae

Genus—Scenedesmus

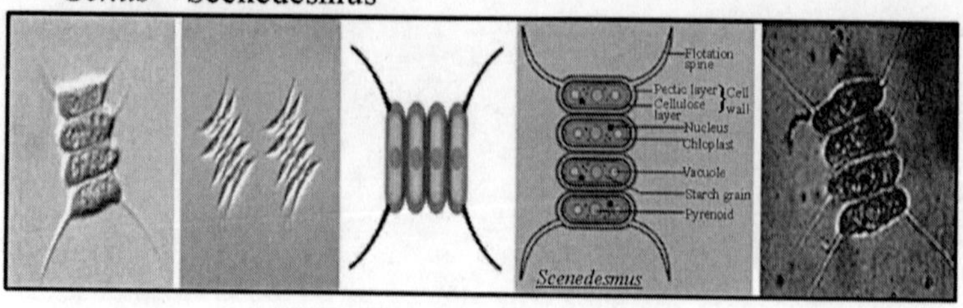

Fig. A Colony of Scenedesmus

1. A fresh water colonial green alga. Unlike most of the colonial green algae that form long filaments, Scenedesmus forms small chains of four cells. The colony is a flat, curved, row of ellipsoidal to fusiform cells arranged in single series with long axis parallel to one another. Number of cells in a coenobium is always a multiple of two, usually 4 or 8.
2. The cell wall is smooth, granulate or spicate, corrugated with or without terminal or lateral teeth or spines, each cell is obtuse, truncated cell poles differentiated by the presence of spines, cell wall made up of pectin.
3. Different forms of coenobia are found including linear, costulatoid, irregular, alternating, or dactylococcoid patterns.
4. Each cell has a single nucleus and chloroplast, laminate with pyrenoid, central vacuole filled with cytoplasm. Small vacuoles, filled with oil, can be seen at either end of each cell. The oil

enables them to float in the water, maximising their exposure to sunlight.

5. Asexual mode of reproduction is seen. Sexual reproduction has been reported only in the laboratory, although flagellated cells (gametes) have been observed in outdoor fish ponds.

(e) Coleochaete

Systematic Position

Class—Chlorophyceae

Order—Chaetophorales

Family—Coleochaetaceae

Genus—Coleochaete

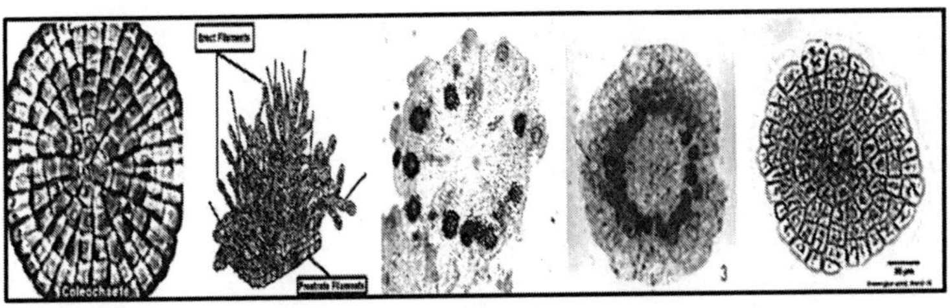

Fig. Coleochaete –Heterotrichous Habit

1. Widely distributed fresh water alga, heterotrichous in habit, both erect and prostrate forms were present. It characterised by true multicellular organisation creating planar discs on solid surfaces in freshwater streams, usually as epiphytes on aquatic plants or growing on the surface of stones.
2. The two forms are cushioned and discoid in the cushioned form the cells are joined end-to end-to form filaments which are prostrate, later become erect.
3. In the discoid forms, all the filaments are prostrate with distinct branches, laterally opposed to form pseudo-parenchymatous disc. Some cells possess long unbranched cytoplasmic seta.
4. The cells are un-inucleate with large parietal chloroplast in which two pyrenoids are embedded.

5. Asexual reproduction takes place with the help of zoospores, aplanospores. Sexual reproduction is oogamous. Oogonia and antheridia are present.

(f) Oedogonium

Systematic Position

Class—Chlorophyceae

Order—Oedogoniales

Family—Oedogoniaceae

Genus—Oedogonium

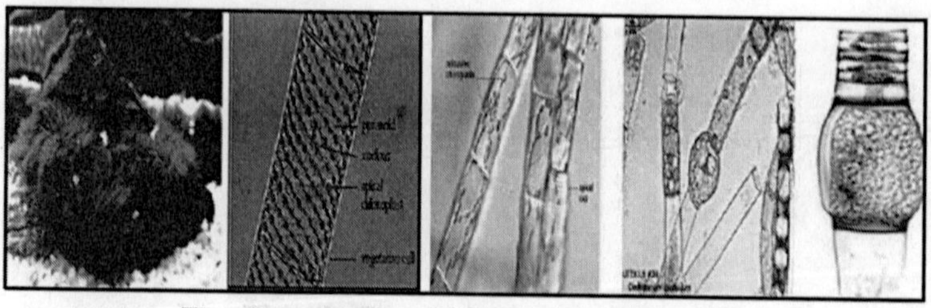

Fig. Organisation of Thallus in Oedogonium

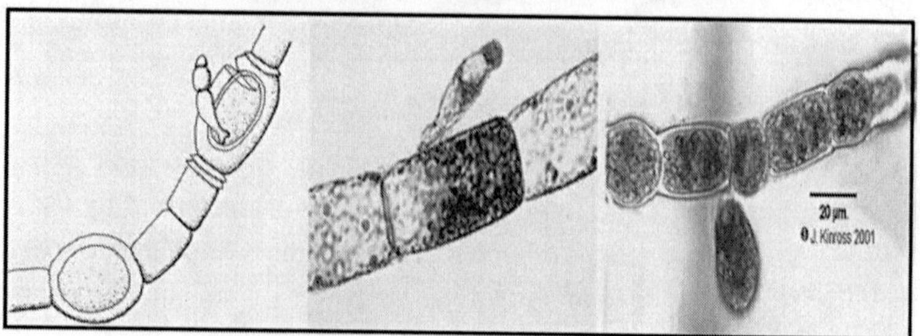

Fig. Nannandrous Species of Oedogonium

1. Oedogonium is a filamentous green algae, with unbranched filaments that are one cell thick. Oedogonium can be free-floating, though it is usually attached to aquatic plants by a holdfast. It appears greenish and inhabits calm, fresh water.
2. Oedogonium, easily identified by the distinctive rings at the apical ends of certain cells. Each cellular division creates a new ring on the cap cell. The cells are cylindrical, sometimes broader at one end, and contain a parietal, net like chloroplast with several pyrenoids. The cell walls are very hard, which makes them an ideal

substrate for epiphytes. Outermost layer made up of chitin, middle pectose and inner one of cellulose.

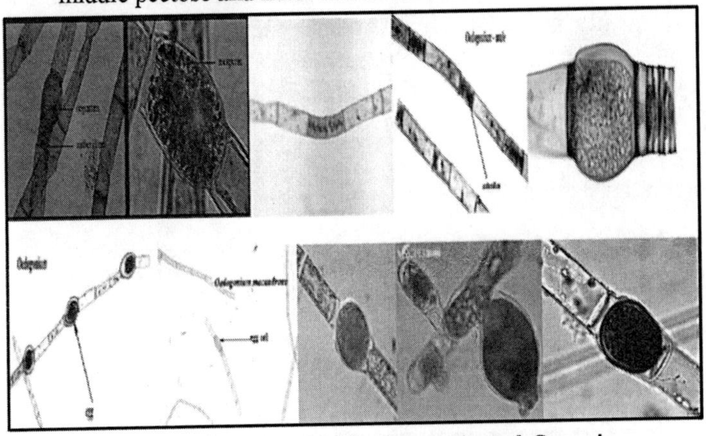

Fig. Filament Bearing Androsporangia and Oogonia

3. Reproduction by vegetative, asexual and sexual methods. Fragmentation is the only vegetative mode.
4. Oedogonium reproduces asexually by non-motile spores and zoospores, which have many flagella. These develop in a zoosporangium cell, one zoospore per zoosporangium. After settling and losing its flagella, a zoospore grows into a filament.
5. Sexual reproduction is oogamous. Antheridia which produce sperm, and oogonia which produce an egg, release the sperm and egg. The egg and sperm then fuse and form a zygote. There are two types of species,
   1. *Macrandous* – In these species, antheridia are borne on filaments of normal size. This group is further subdivided into: Macrandous Monoecious – In these species, antheridia and oogonia are found on the same filament. Macrandous Dioecious – In these species, antheridia and oogonia are borne on different filaments. Although, filaments bearing antheridia and oogonia are morphologically similar, they differ physiologically.
   2. *Nannandrous Species* – In nannandrous species, filaments bearing antheridia and oogonia show morphological distinction. The male filament, which is much smaller than the female filament, is called a dwarf male or nannandrium.

Nannandrous species are always dioecious, i.e., antheridia and oogonia are borne on different filaments. The small male filaments are likely to be attached to a female filament, near an oogonium.

(g) Caulerpa

Systematic Position

Class—Chlorophyceae

Order—Siphonales

Family—Caulerpaceae

Genus—Caulerpa

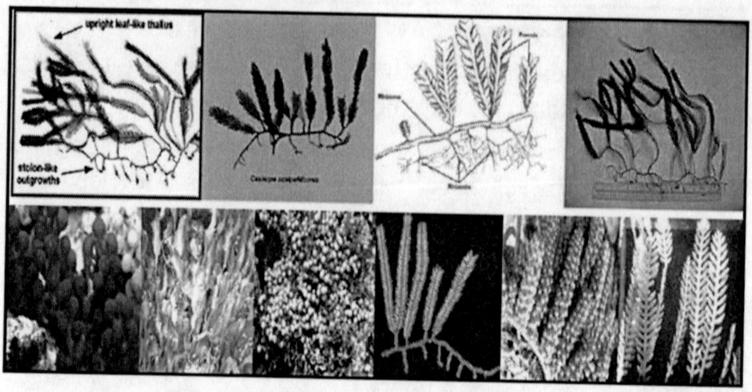

Fig. Organisation of thallus in caulerpa

1. It is a marine macro form restricted to warm seas, shallow deep waters, they are unusual because they consist of only one cell with many nuclei, making them the biggest single cells in the world (coenocytic).
2. The thallus of Caulerpa has rhizome like portion and the rootlet like rhizoids, which vary from species to species. Macroalga with upright leaf-like fronds arising from creeping stolons. The fronds are compressed laterally and the small side branchlets are constricted at the base, where they attach to the mid-rib of each frond.

3. The thallus of *Caulerpa* is a single polynuclear diploid cell, without transverse walls, but interlaced with numerous longitudinal thread like trabeculae (siphon structure) rich in callose and pectin.

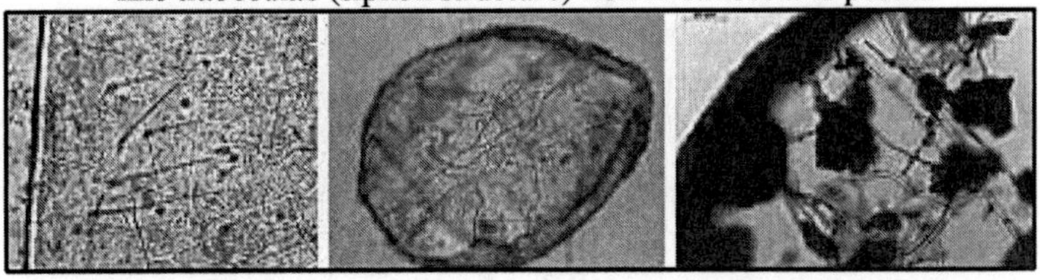

Fig. A Section of the Rhizome of Caulerpa Showing Trabeculae

4. Each cell has multilayered cytoplasm, disciform chloroplast without pyrenoids.
5. Reproduction vegetatively, asexually and sexually. Asexual reproduction by biflagellate zooids, sexual reproduction anisogamous. Propagated by runners.

(h) Chara

Systematic Position

Class—Chlorophyceae
Order—Charales
Family—Characeae
Genus—Chara

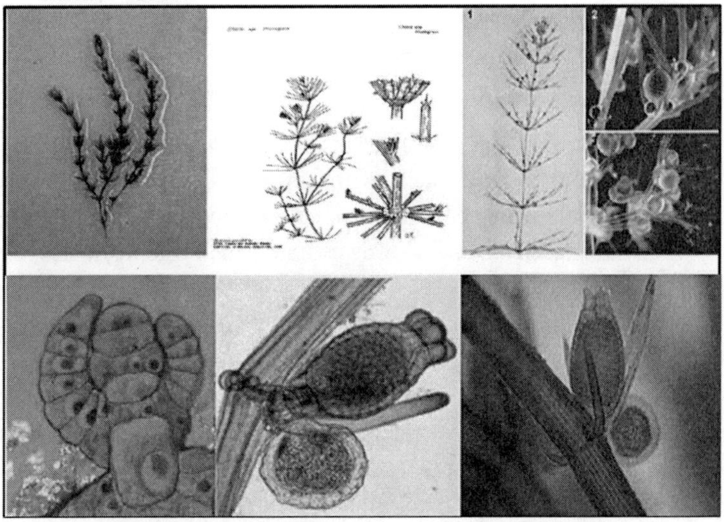

Fig. Chara Entire Plant, Apex, Globule and Nucule

1. It is widely distributed, fresh water algae, plant body made up of erect branches having nodes and internodes, attached to substratum by rhizoids, they are multicellular help in anchorage.
2. At the nodes are present 'leaves' or branches of limited growth, arise other branches from their axil which continue further growth. Single apical cell forms the branches of unlimited growth.
3. The cell structure varies according to position and age, cell wall made up of cellulose and calcium carbonate. Dense cytoplasm with single nucleus, small discoid chloroplasts. Vacuole present.
4. Reproduction vegetative and sexual. Vegetative reproduction is by amylum stars, bulbils, protonema. Sexual reproduction is oogamous.
5. Male sex organ is known as globule, spherical, yellow or red in colour, Female sex organ is nucule, oval in shape, a crown of cells called as corona is present all around its body.

(II) Phaeophyceae

(a) Sargassum

Systematic Position

Class—Phaeophyceae

Order—Fucales

Family—Sargassaceae

Genus—Sargassum

1. Sargassum is a genus of brown macroalgae, species are distributed throughout the temperate and tropical oceans of the world, where they generally inhabit shallow water and coral reefs. The Atlantic Ocean's Sargasso Sea was named after the algae, as it hosts a large amount of Sargassum. Early stages lithophytes later became free floating.
2. It grows to a length of several metres, consist of a holdfast, a stipe, and frond. The lateral branches on the upper side have berry like gas-filled bladders, which help in keeping the fronds afloat and promote photosynthesis. Many have a rough sticky texture, which helps it withstand strong water currents.

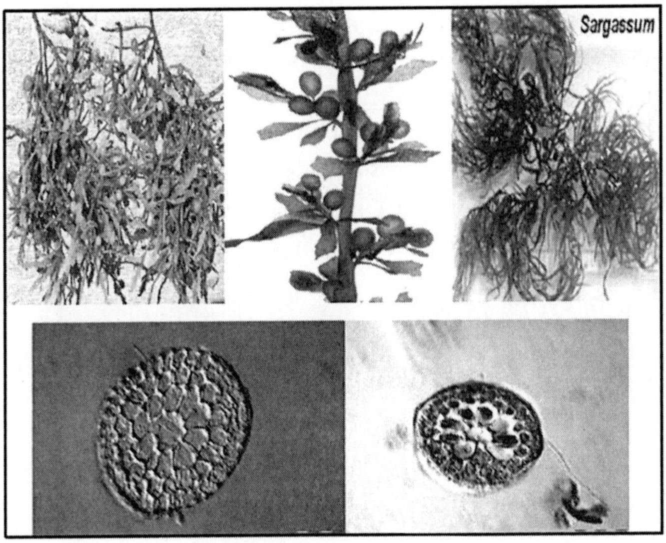

Fig. Entire Plant with Air Bladder, Section of Thallus

3. A section of axis shows single layer of epidermis or meristoderm, middle cortex polygonal cells with intercellular spaces and central thick walled medulla cells without spaces, compact.

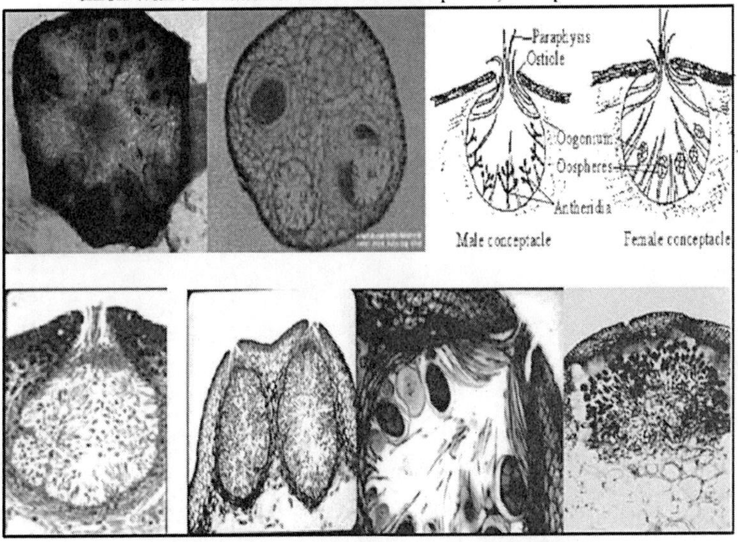

Fig. Section of Male and Female Conceptacle

4. Oogonia and antheridia occur in conceptacles embedded in receptacles on special branches, receptacles terete, shortly

stipitate, arising from the leaf axils, occasionally forked, 10-12 mm long and 1-2 mm in diameter, plants monoecious, male and female reproductive organs in separate conceptacles, but androgenous, with both sexes present in the same receptacles. Ostiole is lined by many hair like sterile structures called as paraphysis.

5. Vegetative reproduction by fragmentation, asexual mode of reproduction absent but sexual reproduction oogamous.

(III) Bacilllariophyceae

(a) Navicula

Systematic Position

Class—Bacillariophyceae

Order—Pennales

Family—Pennaceae

Genus—Navicula

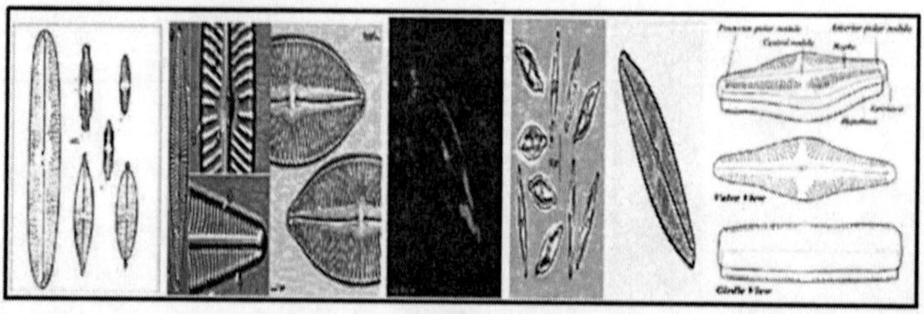

Fig. Frustules of Navicula

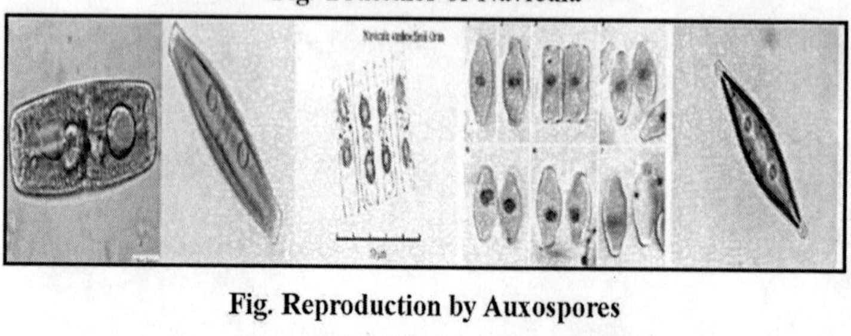

Fig. Reproduction by Auxospores

1. Occurs in standing and fresh waters, the most common type of phytoplankton, cells unicellular elongate, elliptical can exist as colonies. Diatoms are producers within the food chain. A unique feature of diatom cells is that they are enclosed within a cell wall

made of silica (hydrated silicon dioxide) called a frustule.

2. The wall consists of two halves fitting into each other like a box, its overlapping with lid. The outer valve is called as epitheca and the inner one as hypotheca. The margin of the girdle is known as mantle.
3. Chloroplast parietal, green to brown, xanthophylls is present in addition to chlorophyll. Pyrenoids are present in each chloroplast. Photosynthetic produces a fatty oil and volutin.
4. Nucleus lies in a bridge like mass of cytoplasm extending across the centre.
5. Reproduction by cell division or by auxospores.

(IV) Rhodophyceae

(a) Batrachospermum

Systematic Position

Class—Florideae

Order—Nemalionales

Family—Batrachospermaceae

Genus—Batrachospermum

Fig. A Portion of Glomerule Showing Cell Structure

1. Grows in fresh water commonly called as frog spawn algae, prefers slow running streams and moving brooks.
2. The plant body is branched, filamentous, soft and gelatinous. The main axis or central filament has a single row of cells with apical growth called as branches of unlimited growth, lateral branches

from nodal regions, limited growth , short in size called as branches of limited growth. Laterals form clusters at short intervals and one called as glomerule.

3. Each cell of limited and unlimited growth is uninucleate, many chromatophores, single nucleus and a large vacuole.
4. Asexual reproduction is by monospores produced at the end of short branches.
5. Sexual reproduction is oogamous, male reproductive structure is called as antheridium and carpogonium remonius the female one.

(b) Polysiphonia

Systematic Position

Class—Florideae

Order—Ceramiales

Family—Rhodomelaceae

Genus—Polysiphonia

Fig. Thick Tufts of Polysiphonous Thallus

1. The plant forms thick tufts attached to substratum by unseptate rhizoids, filaments placed parallel to each other appear large and elongated, siphonous thallus, hence the name polysiphonia.
2. Other side of central siphon are situated at pericentral siphons, narrow in form, cut small branches from upper ends forming branches of limited growth, called as trichoblasts.
3. The cells are connected by thin cytoplasmic strands, each cell has a single nucleus, chromatophores and floridean starch.

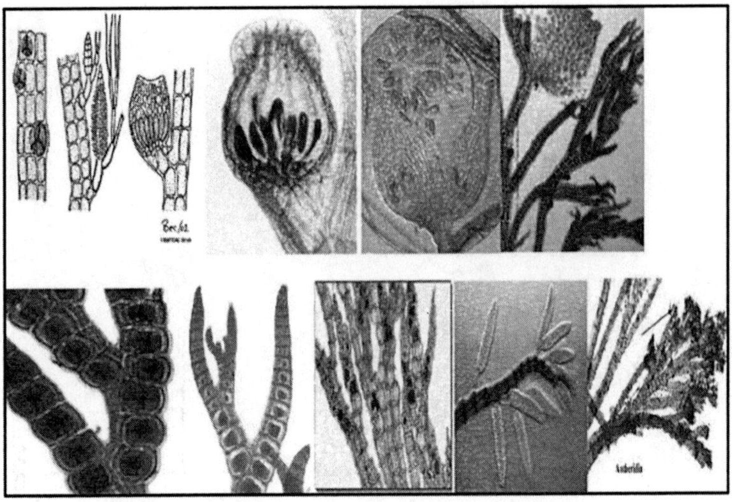

Fig. Thallus Showing Carpospores, Tetraspores and Antheridia

4. Sexual reproduction is oogamous, the male and female reproductive structures known as antheridia and carpogonia respectively. After fertilisation, the cystocarp forms carpospores it is, hence called as carposporangium.
5. Later tetraspores are formed within the tetrasporangium, where four tetra spores are arranged tetrahedrally. The life cycle is triphasic with sex organs, carpospores and tetraspores.

(c) Gracilaria

Systematic Position

Class—Florideae

Order—Gigartinales

Family—Gracilariaceae

Genus—Gracilaria

1. The genus of Gracilaria is cosmopolitan in distribution. It has been reported from the arctic, temperate and tropical regions. The life cycle is triphasic with gametophyte, carposporophyte and tetrasporophyte.

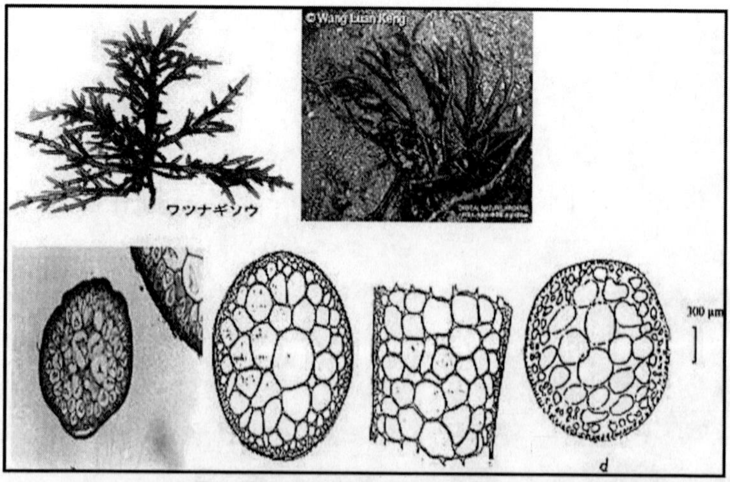

Fig. Habit and Section of Thallus

2. Erect thallus arises from a small discoid holdfast. The thalli are generally cylindrical, depressed or blade-shaped, with lateral, alternate or sub-dichotomous branches. Sometimes, several different branches may be found in one plant. The external appearance of thalli may be used to identify species. The style of the apex and the base of branches are different with species.
3. The vegetative thalli of Gracilaria in cross-section consist of cortex and medulla. Cortical cells are smaller. The outermost 1–2 layers are pigment cells. Medulla comprises large parenchymatous cells. The layers of cortex, the size and number of medullary cells and the change of cells from cortex to medulla are used for identification of species.
4. Reproduction involves antheridia, oogonia or carpogonia. Post-fertilisation, the cystocarp is formed with carposporophyte. The spermatangia are globular or oval, scattered over the surface of thalli. The location and type of spermatangial conceptacle are important considerations for identification of species.

Cystocarp is prominent, protruding, globose or hemispherical, scattered over the surface of the fronds. It may be divided into four portions.

(a) Pericarp consisting of several layers of cells, outmost layer is composed of pigmented cells.

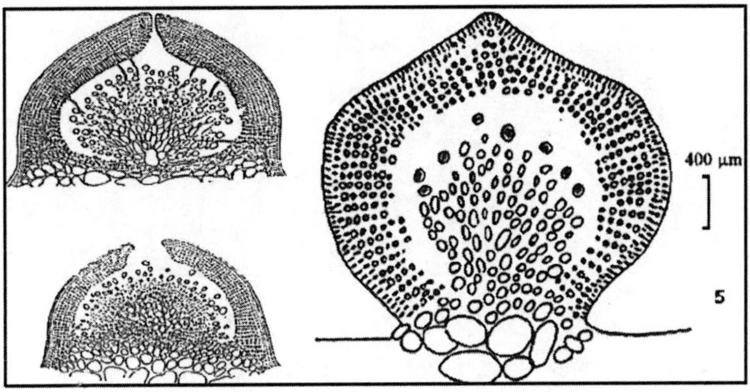

Fig. Section of Thallus with Cystocarp

(b) Gonimoblast in the center of cystocarps consists of parenchymatous cells.

(c) Carposporangia formed at the top of the gonimoblast, round or ovoid in shape.

(d) Absorbing filaments extended from the gonimoblast tissue to the pericarpic-layer. Some species possess absorbing filaments.

5. The tetrasporangia are densely scattered on the cortex surface. Each tetrasporangia is composed of four tetraspores arranged in a cruciate manner. Mature tetrasporophytes (2n) produce (meioses) tetraspores, some of which develop to male gametophytes (n), while others grow to female gametophytes. The male mature gametophyte forms spermatangia, the female forms carpogonia. After fertilisation and cystocarps are formed on the female plant, carpospores (2n) are released and develop to tetrasporophytes again.

(V) Cyanophyceae/Myxophyceae

(a) Oscillatoria

Systematic Position

Class—Myxophyceae

Order—Nostocales

Family—Oscillatoriaceae

Genus—Oscillatoria

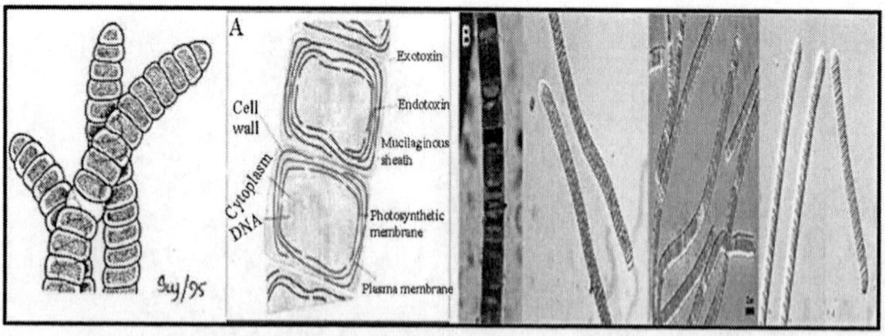

Fig. Trichomes of Oscillatoria

1. Fresh water algae, is found in pools and ponds, unbranched, trichomes occurs in masses. Oscillatoria is a genus of filamentous cyanobacterium, which is named for the oscillation in its movement.
2. Each trichome is cylindrical cells placed end to end, apical cell rounded covered by mucilaginous sheath. The tip of the trichome oscillates like a pendulum.
3. Each cell has a central colourless centroplasm or incipient nucleus and peripheral chromoplasm, prokaryotic in nature. Moves in oscillating fashion hence the name oscillatoria.
4. Vegetative reproduction by hormogonia, fragmentation. Oscillatoria forms long filaments of cells, which can break into fragments called hormogonia. The hormogonia can grow into a new, longer filament. Breaks in the filament usually occur where dead cells, necridia are present.
5. Asexual and sexual modes of reproduction is absent.

(b) Nostoc

Systematic Position

Class—Myxophyceae

Order—Nostocales

Family—Nostocaceae

Genus—Nostoc

1. Nostoc, a terrestrial or fresh water algae found free floating or attached to substratum also endophytic within Anthoceros.

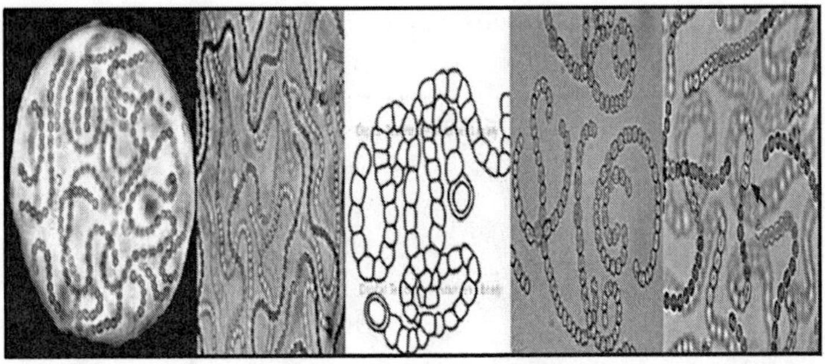

Fig. Filaments of Nostoc

2. Cells arranged in beadlike chains that are grouped together in a gelatinous mass, mucilaginous sheath firm. Each cell is spherical monoliform in appearance.
3. Like most blue-green algae, *Nostoc* contains two pigments, blue phycocyanin and red phycoerythrin, as well as chlorophyll, and has the ability to fix nitrogen in specialised cells called heterocysts.
4. The protoplast is easily distinguishable into peripheral pigmented portion, chromoplasm and central colourless region, centroplasm. Gas vesicles occur in the vegetative cell, buoyant enough to be planktonic.
5. Vegetative reproduction is by fragmentation, hormogones, heterocysts and akinete. A special thick-walled cell akinete has the ability to withstand desiccation for long periods of time and helps in formation of new filaments.

(c) Anabaena

Systematic Position

Class—Myxophyceae

Order—Nostocales

Family—Nostocaceae

Genus—Anabaena

1. Fresh water algae, found in pools and ponds, unbranched , trichomes occurs in masses, endophytic in roots of

Cycas, fronds of Azolla and Anthoceros, lower part of the thallus.

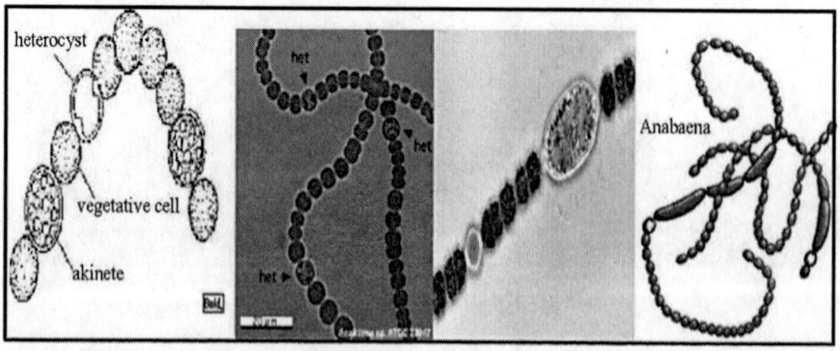

Fig. Filament of Anabaena Showing Akinete and Heterocyst

2. Cells Placed end to end, Spherical to barrel shaped numerous puleracalasy heterocyysts present, have akinete may.
3. Each cell has a central colourless centroplasm or incipient nucleus and peripheral chromoplasm, is prokaryotic in nature.
4. Characteristic feature of Anabaena, it forms a watery gelatinous sheath, never a firm colony. The trichomes are never contorted, usually hyaline.
5. Vegetative reproduction is by fragmentation, akinetes is conspicuous always larger than vegetative cells, cylindrical with rounded edges. Asexual and sexual modes of reproduction are absent.

II FUNGI

(I) Phycomycetes

Oomycetes

(a) Albugo/Cystopus

Systematic Position

Class—Oomycetes

Order—Peronosporales

Family—Albuginaceae

Genus—Albugo

1. It is an obligate parasite on angiospermous plants causing leaf spot disease. This organism causes white rust or white blister diseases in above-ground plant tissues. Symptoms of white rust caused by Albugo typically include yellow lesions of uneven shape and sizes, on the upper leaf surface and white pustules on the underside of the leaf.

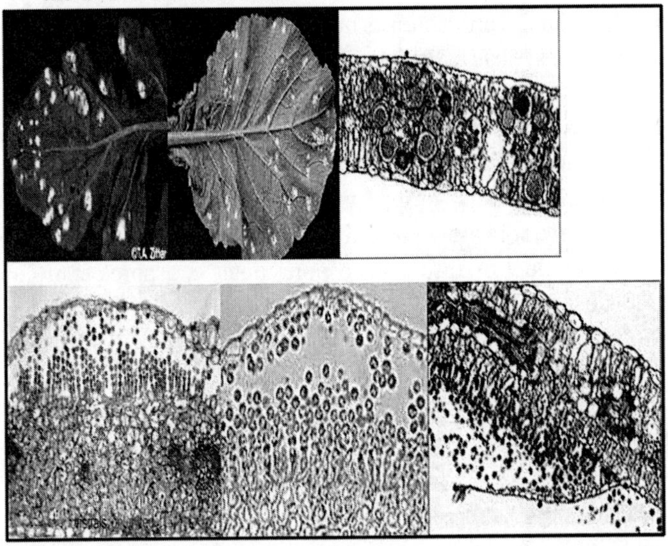

Fig. White Rust of Albugo, Infected Leaf, Conidia And Sex Organs

2. Mycelium, intercellular, coenocytic, multinucleate and branched, producing knob like intracellular haustoria. Cell wall made up of chitin and cellulose, fund vaculated cytoplasm with small nuclei.
3. The Albuginaceae reproduce by producing both sexual spores called oospores and asexual spores called sporangia in a many-stage polycyclic disease cycle. The infected portion shows several erect conidiophores, in basipetalous succession. Sexual reproduction is oogamous, sex organs are deeply located within host tissue. Consequently, appearance of external symptoms like hypertrophy and distortion.
4. The asexual reproduction takes place by means of biflagellate zoospores inside the sporangia, the hyphae accumulate just beneath the epidermis of the infected leaf. Hyphae, certain thick-walled, clavate aerial sporangiophores come out. Successively

the sporangia develop by constriction method, in basigenous chains, oldest conidium at the top and youngest at the base is gelatinous pad develops in between two sporangia acting as a separator of the two sporangia from each other. Each conidium is polyhedral or globose and multinucleate, biflagellate zoospores are produced by the conidia.

5. The sexual reproduction is oogamous. The sex organs develop on the hyphal ends in the intercellular spaces of the deeper tissues of the host. The female sex organs are oogonia and the male sex organs are antheridia. The oogonium is rounded, outer multinucleate periplasm and inner uninucleate ooplasm, while the antheridium is club-shaped, multinucleate. The developing oogonia and antheridia are separated from the rest of mycelium by septa. The product of fusion is oospore, outer exospores spiny inner endospore thin.

(II) Zygomycetes

(b) Mucor

Systematic Position

Class—Zygomycetes

Order—Mucorales

Family—Mucoraceae

Genus—Mucor

1. Mucor is saprophytic fungus and grows on dead organic material. Commonly found in soil, digestive systems, plant surfaces, and rotten vegetable matter, it is known as black mould. The mycelium looks like fine cotton threads. After formation of spores, it changes into a black mass called black mold fungus.
2. Mycelium made up of coenocytic hyphae, multinucleate with vacuolated granular cytoplasm, branched cottony. Cell wall rich in chitin, haustoria absent. Hyphae with glycogen and oil drops, embedded in granular cytoplasm.

3. Three modes of reproduction are present. Vegetative reproduction: It takes place by fragmentation of hyphae. In Mucor, the asexual reproduction may take place either by thick walled intercalary chlamydospores or by formation of oidia. In favourable condition mycelium gives out some aerial hyphae known as sporangiophore, tip of the sporangiophore swells to form a globose sporangium with uninucleate, haploid sporangiospores. Extension of the sporangiophore called the columella protrudes into the sporangium. On maturity, the outer wall of sporangium bursts and the spores are dispersed by wind. Each spore on getting suitable conditions germinate by producing germ tubes which develops into new mycelium.

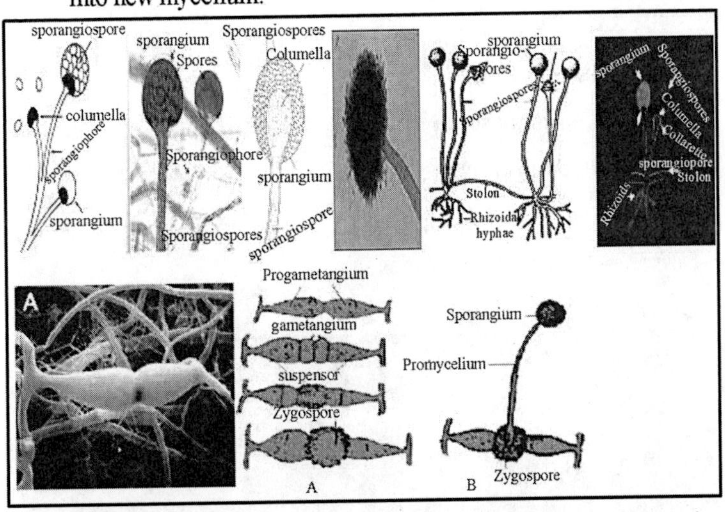

Fig. Sporangium of Mucor, Fusion of Hyphae, Gametangium Formation

4. It is an heterothallic, dioecious form. Isogamous type of sexual reproduction. Male and female mycelia are identical but physiologically different, thus represented by positive and negative strains. During sexual reproduction, two hyphae of different strains grow towards each other. Such hyphae are called progametangia. Their tips swell up and a transverse septum develops in each differentiating a terminal cell or gametangium. Lower portion is called suspensor. Gametangia of strains are multinucleate and gametangium is hence called coenogamete.

These two coengomaetes fuse together, zygote or zygospore is formed.

5. Zygote develops a thick wall of two layers, the outer layer is called as exospore and inner endospore. The zygote can tide over unfavourable environmental condition. At the time of germination, there is production of zygosporangiophore terminating into multinucleate zygosporangium.

(II) Ascomycetes

(a) Saccharomyces

Systematic Position

Class—Ascomycetes

Sub Class—Hemiascomycetes

Series—Plectomycetes

Order—Endomycetales

Family—Saccharomycetaceae

Genus—Saccharomyces

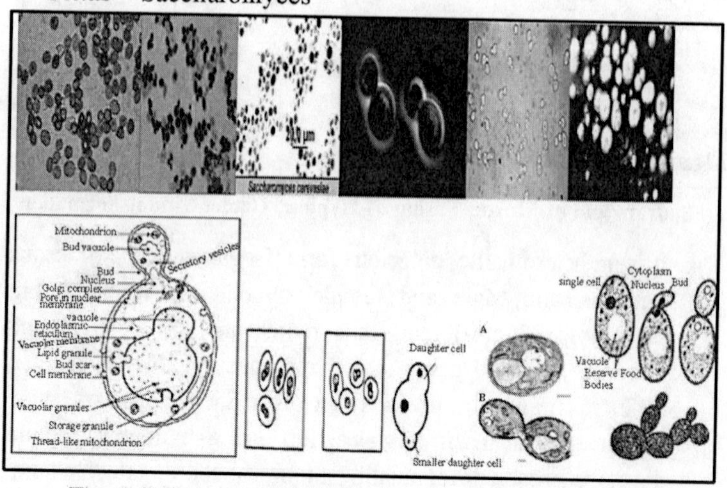

Fig. Cell Structure and Methods of Reproduction in Yeast

1. Yeast is a microscopic unicellular, uninucleate fungus varying from 2-25 micron in length and 1-5 micron in width. Yeasts are

eukaryotic microorganisms. Shape ranges from spherical to oval to ellipsoidal. The term "yeast" is often taken as a synonym for *Saccharomyces cerevisiae.*

2. Cell is enveloped by a cell wall, composed of two polysaccharides, glucan and mannan, proteins, lipids and chitin. Inner to the cell wall, there occurs lipoproteinaceous cell membrane, surrounding the protoplasm, mitochondria, golgi apparatus.
3. The reserve food is glycogen, oil globules and volutin granules. A large central vacuole containing hydrolytic enzymes occurs at the centre of the cell. The nucleus is distinct being enveloped by two unit membranes and perforated by pores, primitive but complicated type.
4. Yeasts has vegetative, asexual and sexual reproductive cycles. The common mode of vegetative reproduction is budding. Here, a small bud known as a bleb, or daughter cell, is formed on the parent cell. The nucleus of the parent cell splits into a daughter nucleus and migrates into the daughter cell. The bud continues to grow until it separates from the parent cell, forming a new cell. Fission also seen, creating two identically sized daughter cells.
4. Asexually takes place by enlarged yeast cell forming the 'ascus.' The nuclues of the ascus divides twice, four nuclei and the four ascospores are formed. Sometimes eight ascospores may also be produced. Each ascospore is surrounded by a thick wall. These spores are perennating bodies. They remain dormant in adverse conditions. Favourable conditions they germinate, ascospores liberate in the atmosphere, germinate and new individuals are formed by budding.
5. Sexual reproduction isogamous, takes place by conjugation. Two individual cells come close to each other and the beak-like outgrowth is formed. These outgrowth fuse with each other. The nuclei of individuals come in these beaks, the wall of contact dissolves and ultimately the nuclei fuse with each other giving rise to a zygote, which soon converts into an ascus, divides thrice producing eight nuclei. The first division is reductional and bring again haploid condition.

(b)Aspergillus

Systematic Position

Class—Ascomycetes

Sub Class—Euascomycetidae

Series—Plectomycetes

Order—Eurotiales

Family—Eurotiaceae

Genus—Aspergillus

Fig. Thallus Organisation in Aspergillus: Black Mold

1. A saprophytic fungus, not yet found a host, non-pathogenic species, obtain nutrients from soil, plant detritus, or wood but pathogenic, opportunistic organism, obtains nutrients from a host. Aspergillus, gets its name from a vessel in the shape of a circle, with filamentous extension growing out. This resembles the shape of an aspergillum, a device used for sprinkling holy water.
2. Hyphae constitute the mycelium, septate, hyaline, branched and uninucleate. Wall composed of chitin and fungal cellulose. Cytoplasm is granular, inclusions are like oil drops.

3. Aspergillus species can reproduce sexually and asexually, although asexual reproduction predominates in most species, mostly consist of dense felt of erect, non-septate, unbranched conidiophores, terminate in vesicle covered with either a single palisade-like layer of phialides, uniseriate or a layer of metulae which bear small whorls of phialides, biseriate. The vesicle, phialides, metulae and conidia form the conidial head. Conidia are one celled, smooth walled, hyaline or pigmented and are basocatenate, forming long dry chains, which may be divergent arranged basipetally. Some species may produce sclerotia.
5. Sexual reproduction takes place between sexually compatible strains belonging to different groups. In the sexual cycle, haploid hyphae connect in a process called plasmogamy. First, the plasma membranes fuses, in a dikaryon stage, nuclei fuse, karyogamy and the cell membrane shapes into an ascus, an elongated, oval shaped cell.
6. All the ascus are protected by ascocarp or fruiting body, closed, rounded cleistothecium lined by interwoven hyphae constituting peridium. This diploid ascus meiotically divides, forming four biconvex un-inucleate, ascospores, which are released from the ascocarp.

(c) Penicillium

Systematic Position

Class—Ascomycetes

Sub Class—Euascomycetes

Series—Plectomycetes

Order—Eurotiales

Family—Eurotiaceae

Genus—Penicillium

1. A saprophytic fungus commonly called as blue mold, is found on prickles, foods and fruits and of major importance in the natural environment as well as food and drug production.

2. The thallus mycelium consists of highly branched network of multinucleate, septate, usually colourless hyphae. Apical hyphae are un-inucleate; older ones are multi-nucleate. Many-branched conidiophores sprout on the mycelia, bearing individually constricted conidiospores. Cytoplasm are rich in oil globules.

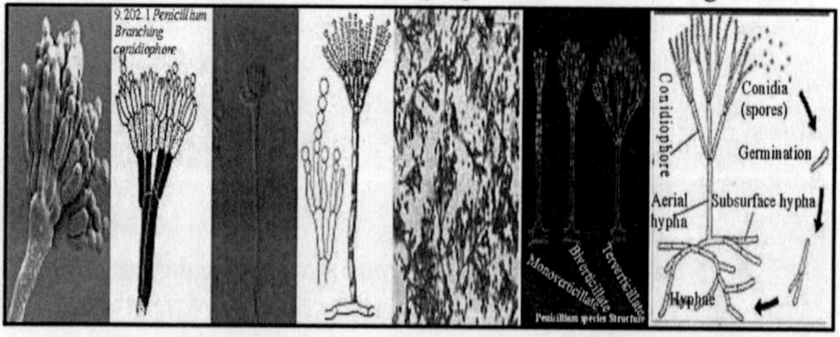

Fig. Thallus Organisation in Peniccilium: Blue Mold

3. Branches of conidiophores are called as metulae, tips bear sterigmata, which later bears secondary sterigmata at the tips of which conidia are formed in chains. Conidia are produced basipetally in chains, each conidia is oval.
4. Sexual reproduction involves the production of ascospores, commencing with the fusion of an archegonium or ascogonia and an antheridium, zygote formation. Fruiting body is closed, cleistothecium, wall two layered and peridium. The irregularly distributed asci contain eight unicellular, pyriform ascospores each.
5. A wide variety of penicillium are the source major antibiotics, Penicillin, a drug produced by *P. chrysogenum,* was accidentally discovered by Alexander Fleming in 1929, and is found to inhibit the growth of gram-positive bacteria.

(III) Basidiomycetes

(I) Heterobasidiomycetidae

(a) Puccinia

Systematic Position

Class—Basidiomycetes

Sub Class—Heterobasidiomycetidae

Series—Plectomycetes

Order—Uredinales

Family—Pucciniaceae

Genus—Puccinia

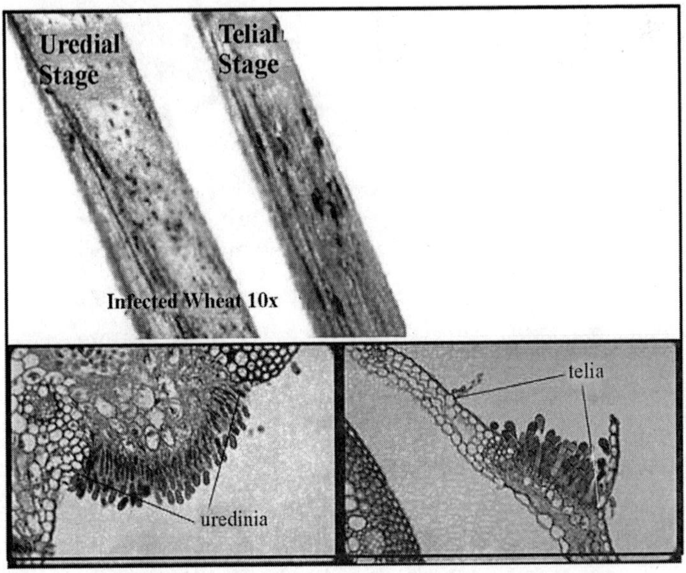

Fig. Wheat Leaf Infected By Uredinia and Telia

1. The black or cereal rusts are caused by the fungus puccinia graminis, significant disease affecting cereal crops. Its life cycle has two hosts, wheat and barberry, heteroecious. Mycelium septate, binucleate and branched. Haustoria) present. Clamp connections are seen. Polymorphic form with uredospores, teleutospores, pycniospores and aeciospores.
2. Stem rust on wheat is characterised by the presence of uredinia on the plant, which are brick-red, elongated, blister-like pustules which are easily shaken off. They frequently occur on the leaf sheaths, but are found on stems, leaves, glumes and awns. Each uridium has several stalked uredospores, oval, double walled, binucleate, exine spiny.These spores on germination give rise to binucleate mycelium.

3. Towards the end of the growing season, black telia are produced, stem rust is also known as black rust. The telia are firmly attached to the plant tissue. The site of infection is a visible symptom of the disease. Teleutospores, stalked, spindle shaped, binucleate with septum, bicelled, smooth exine. These spores form basidium which produces basidiospores.

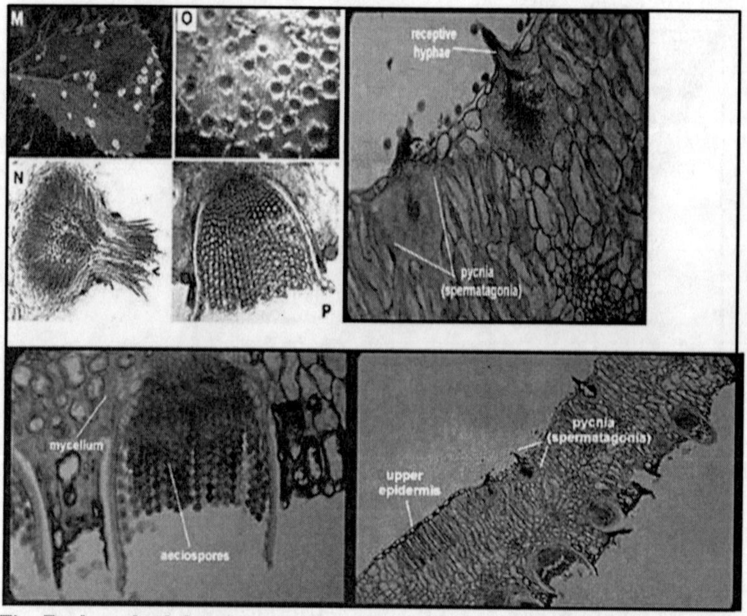

Fig. Barbery leaf showing spermagonia, receptive hyphae and Aeciospores

4. The basidiospores infect the barbery leaf giving rise to pycnidial patches on dorsal side and aecidial on ventral side. Pycnidial cups are globose, ostiole present at the tip, contains slender elongated spermatiophores, from which arise pycniospores or spermatiophores. Ostiole helps in penetration of several receptive hyphae. Pycnia appear on barberry plants in the spring, usually in the upper leaf surface. Five to 10 days later, cup-shaped structures filled with orange-yellow, powdery aeciospores break through the lower leaf surface.
5. The ventral side has aecidial cups facing the earth and enclosed by peridium. Cups remain open, cup connected with intercellular binucleate hyphae. Several binucleate sporophores are present in the cup, parallel, separated from each other by a disjuncter cell.

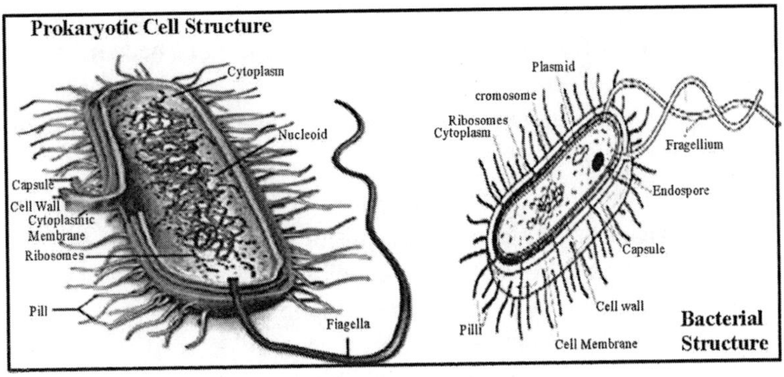

Fig. Ultrastructure of a Bacterial Cell

1. Unicellular, non chlorophyllous, micro- organisms with parasitic, saprophytic or autotrophic mode of nutrition. The bacterial cell wall differs from that of other organisms by the presence of peptidoglycan, which is located immediately outside the cytoplasmic membrane. Peptidoglycan is made up of polysaccharide backbone consisting of alternating N-Acetylmuramic acid (NAM) and N-acetylglucosamine (NAG) residues in equal amounts. Nucleus is not well organised.

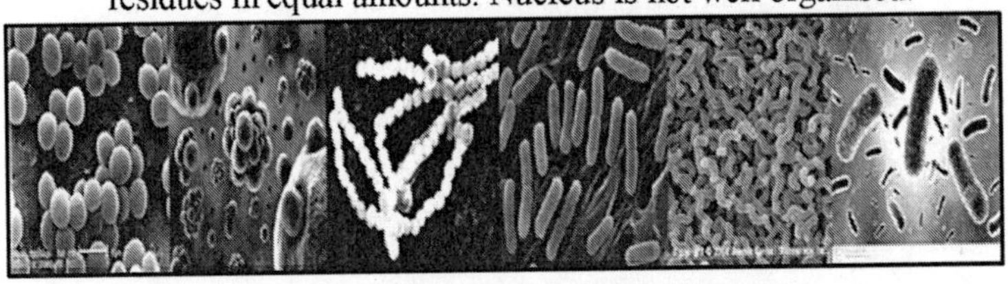

Fig. Various Forms and Shapes of Bacteria

2. Phospholipid bilayer, of inner membrane is impermeable to charged molecules. However, channels called porins are present in the outer membrane that allow for passive transport of many ions, sugars and amino acids across the outer membrane. These molecules are therefore present in the periplasm, the region between the cytoplasmic and outer membranes.
3. Flagella are whip-like structures protruding from the bacterial cell wall and are responsible for bacterial motility. The arrangement of flagella about the bacterial cell is unique to the species. Fimbrae are protein tubes that extend out from the outer membrane, short

in length present in the entire bacterial cell surface. Fimbrae usually function to facilitate the attachment of a bacterium to a surface or to other cells. Pili are involved in the process of bacterial conjugation, where they are called conjugation pili or "sex pili."

4. Cell prokarotic, DNA is not enclosed inside membrane-bound nucleus but instead resides inside the bacterial cytoplasm. This means that the transfer of cellular information through the processes of translation, transcription and DNA replication occur within the compartment and can interact with other cytoplasmic structures, mostly ribosomes. Most bacterial DNA are circular, with chromosomal DNA, also contains small independent pieces of DNA called plasmids that often encode for advantageous traits but not essential to their bacterial host. Plasmids can easily be gained or lost by a bacterium and can be transferred between bacteria as a form of horizontal gene transfer. So, plasmids can be described as extra chromosomal DNA in a bacterial cell. Various shaper seen, coccus (spherical), bacillus (rod-like) spirillum (spiral), vibrio (comma) and filamentous (elongated).
5. Asexual reproduction by binary fission and endospore formation. Sexual by conjugation, transformation and transduction Endospores are bacterial survival structures that are highly resistant to many different types of chemical and environmental stresses and enable the survival of bacteria in environment that would be lethal for the cells in their normal vegetative form.

IV VIRUSES

Systematic Position

Class—Schizomycetes

Order—Virales

Family—Viraceae

Genus—Tobacco mosaic virus

1. Viruses are chemical packages of proteins and nucleic acids. Minute organisms smaller than bacteria. A virus is a small infectious agent that replicates only inside the living cells of other organisms. Viruses can infect all types of life forms, from animals and plants to bacteria and archaea.

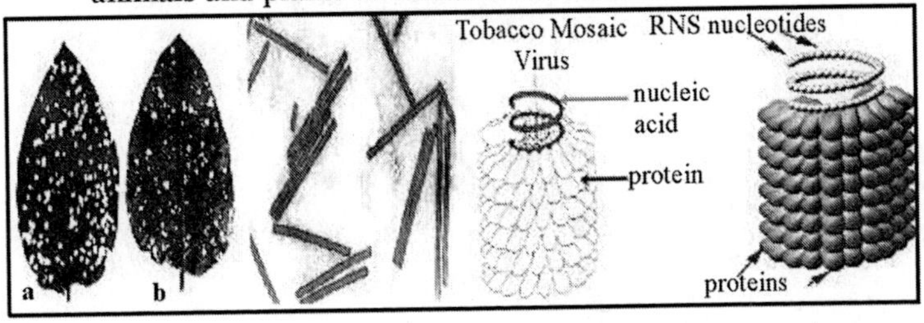

Fig. Infected Leaf, Causal Organism and Ultrastructure of TMV

2. Virus particles known as virions consist of two or three parts: i) the genetic material made from either DNA or RNA, long molecules that carry genetic information; ii) a protein coat that protects these genes; and in some cases iii) an envelope of lipids that surrounds the protein coat, when they are outside a cell. The shapes of viruses range from simple helical and icosahedral forms to more complex structures. Viruses spread in many ways; viruses in plants are often transmitted from plant to plant by insects that feed on plant sap, such as aphids; viruses in animals can be carried by blood-sucking insects.
3. Tobacco mosaic virus (TMV) is a positive-sense single stranded RNA virus that infects plants, especially, infection causes characteristic patterns, such as "mosaic"-like mottling and discoloration on the leaves, TMV was the first virus ever to be discovered.
4. Tobacco mosaic virus has a rod-like appearance. Its capsid is made from 2,130 molecules of coat protein and one molecule of genomic single strand RNA, 6,400 bases long. The coat protein self-assembles into the rod like helical structure (16.3 proteins per helix turn) around the RNA, which forms a hairpin loop structure.
5. Like other plant pathogenic viruses, TMV has a very wide host range and has different effects depending on the host being infected. The tobacco mosaic virus has been known to cause a

production loss for tobacco. The first symptom of this virus disease is a light green coloration between the veins of young leaves, followed by the development of a "mosaic" or mottled pattern of light and dark green areas in the leaves. Rugosity may be seen where the infected plant leaves display small localised random wrinkles.

(V) LICHENS

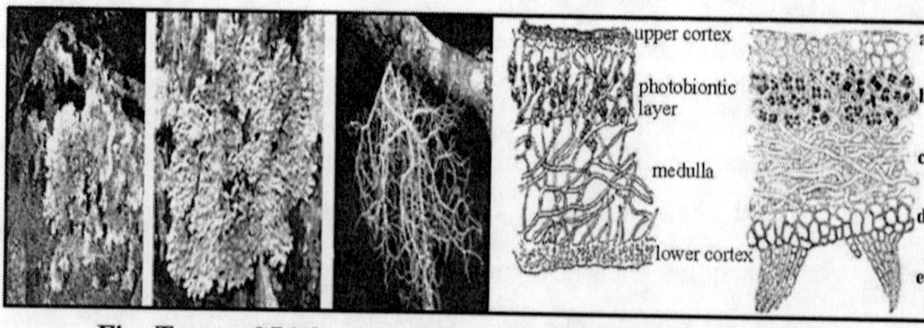

Fig. Types of Lichens and Cross Section of Lichen Thallus

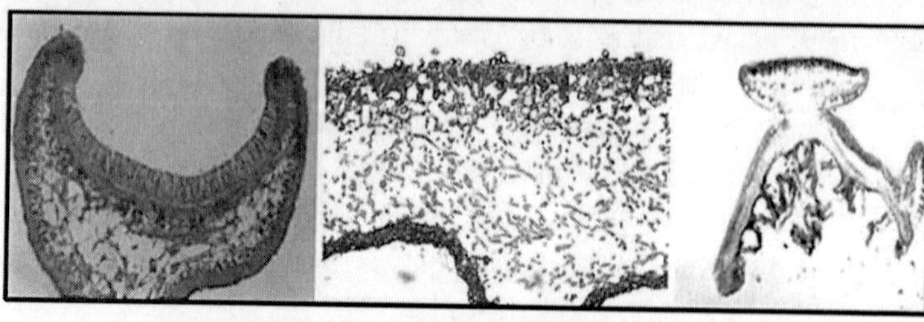

Fig. Cross Section of Fruiting Body of Lichens

1. A lichen is composite organism, an association of a fungus with algae in which the two organisms live intimately to form a single thallus, consisting of a fungus the mycobiont and a photosynthetic partner the photobiont or phycobiont growing together in a symbiotic relationship. The photobiont is usually either a green algae or cyanobacterium. The external morphology, anatomy and physiology of the thallus is different from either of the component.
2. Lichens occur in some of the most extreme environment on Earth, arctic tundra, hot deserts, rocky coasts, and toxic slag heaps. However, they are abundant as epiphytes on leaves and branches in rain forests and temperate woodland, on bare

rock, including walls and gravestones, and on exposed soil surfaces.

3. Section reveals upper cortex, algal zone, medulla and lower cortex. Cortex cell spaces filled with gelatinous material epidermis like, usually single layered. The algal zone with blue green algae or green algae interwoven with fungal hyphae, algal photosynthetic with chlorophyll, multiply by cell division or aplanospores fungal hyphae with haustoria, absorb nutrients. Algae is for assimilation and fungi is for absorption. Beneath the upper cortex is an algal layer composed of algal cells embedded in densely interwoven fungal hyphae.

   Each cell or group of cells of the photobiont is usually individually wrapped by hyphae, and in some cases penetrated by an haustorium. Beneath this algal layer is a third layer of loosely interwoven fungal hyphae without algal cells. This layer is called the medulla.

   Beneath the medulla, the bottom surface resembles with the upper surface and is called the lower cortex, consisting again of densely packed fungal hyphae. The lower cortex often bears root like fungal structures known as rhizines, which serve to attach the thallus to the substrate on which it grows.

4. Three methods of reproduction, vegetative by isidia, soredia. Soredia are small groups of algal cells surrounded by fungal filaments that form in structures called soralia, from which the soredia can be dispersed by wind.

   Another form of diaspore are isidia, elongated outgrowth from the thallus that break off for mechanical dispersal. Asexual is confined to fungal partner with flask shaped fruiting body, pycnidia, opening ostiole, chain of non motile conidia. Conidia is light and dry.

5. Sexual reproduction of two types is based on the fungal partner, ascolichen and basidiolichen, Among the ascolichens, spores are produced in spore-producing bodies or fruiting bodies, the three most common spore body types are the apothecia, perithecia and the pycnidia, while basidiolichen forms basidiocarps.

VI. BRYOPHYTA

(i) Hepaticopsida

(a) Marchantia

Systematic Position

Class—Heptaticae

Order—Marchantiales

Family—Marchantiaceae

Genus—Marchantia

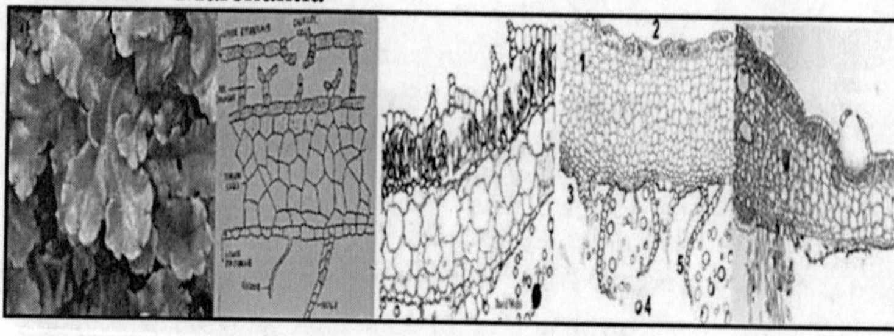

Fig. Section of Thallus of Marchantia

1. Genus is found in cool, damp climate, at the banks of rivers, streams and on rocks.Thallus of Marchantia is dark green dichotomously branched, dorsiventral, distinct mid-rib, an apical notch with wavy margin. Ventral surface on either sides of the midrib two rows of scales are present. Lower side having rhizoids smooth walled and tuberculated, is used for fixation and absorption.
2. The V.S. of the thallus shows upper photosynthetic region and lower storage region. Upper epidermis single layered with barrel shaped pores, air chambers below pores help in gaseous exchange. Chlorophyllous filaments form the assimilatory region, have storage layer without intercellular spaces, full of starch.
3. Dorsal surface has numerous cup like structures, gemma cups. These are vegetative reproductive bodies. Each gemma is a globular multicellular structure intermixed with numerous clubshaped mucilage hairs. The upper portion is discoid and constricted in the middle, cells rich in chloroplast and oil. Favourable conditions develop into new gametophyte.

4. It is a dioecious plant, male gametophores terminated by an umbrella like structure while female gametophores radiating, drooping lobes on a cap.

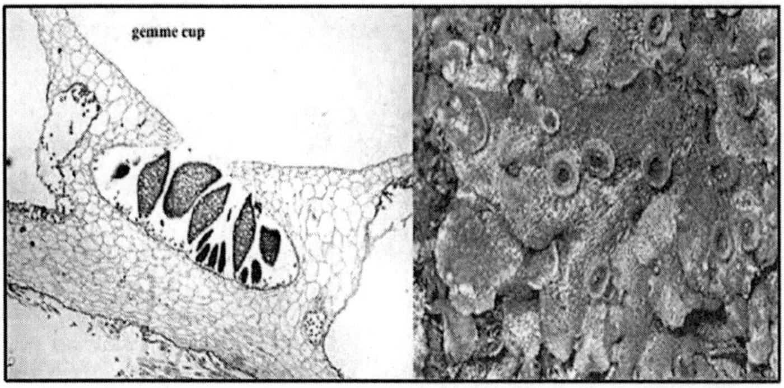

Fig. Section of Thallus of Marchantia Showing Gemma Cups

Sex organs are present at the apices of certain branches of the thallus. The individual antheridia are club shaped with short stalk, sunk in depressions from which the spermatozoids are in some cases forcibly ejected, while the archegonia are protected by disk or by a special two-lipped involucres, perigynium. Each archegonium is inverted flask like structure having a swollen venter and narrow neck. Venter has the egg nucleus.

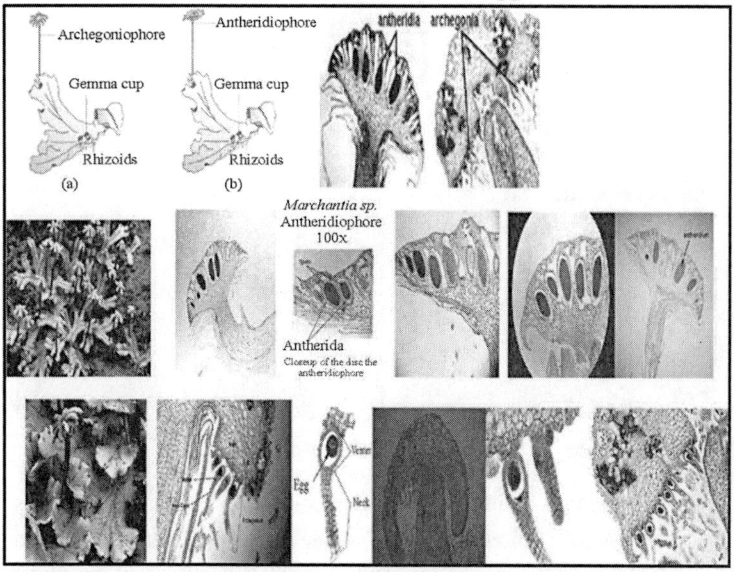

Fig. Section of Gametoph Yte of Marchantia Showing Sex Organs

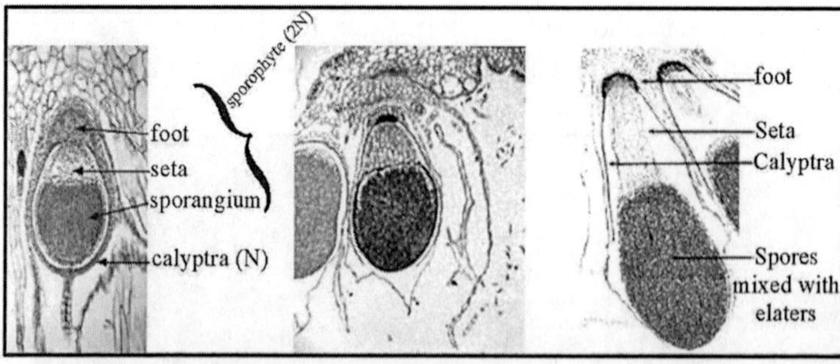

Fig. Section of Sporophyte of Marchantia Showing Spores

5. The sporophyte is differentiated into foot, seta and capsule. The foot is broad, terminates into a long narrow seta later broad oval capsule, several uninucleate spores and elaters are found within. Each spore spherical thick, elaters spindle shaped , long, with spiral thickenings, helps in dispersal of spores.

(ii) Anthocerotopsida

(a) Anthoceros

Systematic Position

Class—Anthocerotopsida

Order—Anthocerotales

Family—Anthocerotaceae

Genus—Anthoceros

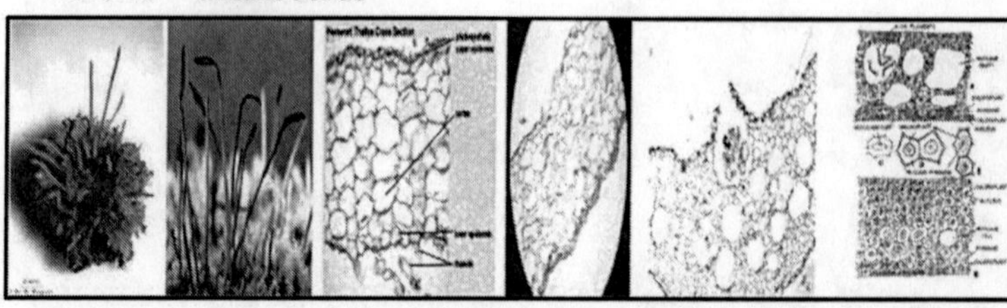

Fig. Section of Gametophytic of Anthoceros

1. Common in damp localities, is close to waterfalls. Thallus dorsiventral, prostrate, translucent, no distinct midrib, margin crenate, thallus lobed. Ventral side is with numerous rhizoids.

Scales absent. Algal colonies appear as dark thick spots, on ventral side of thallus.

2. Section of the thallus shows homogenous, parenchymatous, polygonal cells without intercellular spaces. Air pores are present but air chambers are absent. Scattered in this homogenous mass of cells cavities are present with Nostoc filaments. Upper epidermis continuous, non-chlorophyllous. Each cell has a chloroplast with single pyrenoid and nucleus.

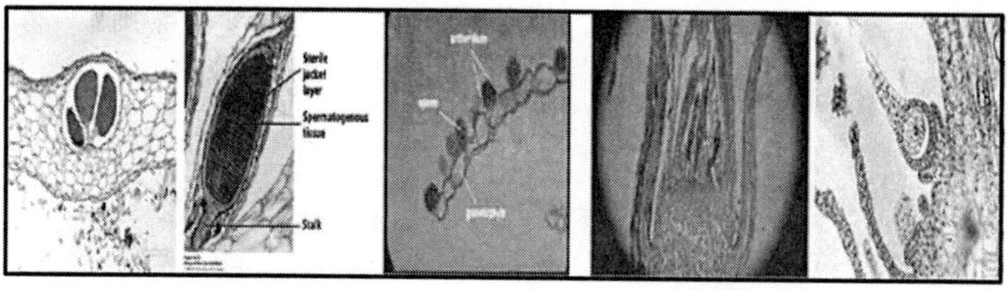

Fig. Section of Gametophyte of Anthoceros Showing Sex Organs

3. The thallus may be monoecious or dioecious, sex organs embedded in the dorsal part of the thallus. Male reproductive structures antheridia, stalked and branched, jacket surrounds the club shaped antheridia, androcytes many on division give rise to single linear, biflagellate antherozoid.

4. Archegonium is the female reproductive structure. Well protected by a chamber, it is distinguished into two zones swollen venter, narrow neck. The base swollen to form a venter, having the egg. The venter canal cell short lived, neck canal cells form a single row. After fertilization the sporophyte is attached to the gametophyte.

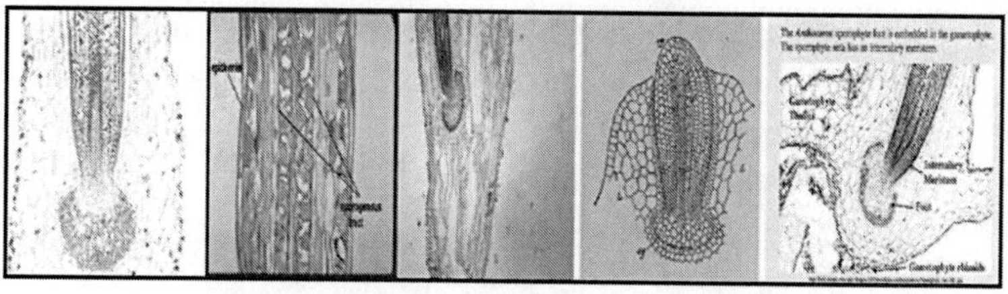

Fig. Section of Sporophyte of Anthoceros Showing Spores

5. The sporophyte consists of a large bulbous foot, embedded in the superficial cells of gametophyte, which grows into a long capsule, rich in meristematic cells, attains one inch and a half length.

The capsule wall is thick, and the epidermis contains functional stomata and cells possess chlorophyll, it is capable of assimilation, central region of the capsule has a strand of narrow elongated thick cells forming the columella, and spores mixed with elaters are formed from the dome-shaped archesporiumbetween this and the wall.

(iii) Musci

(a) Polytrichum

Systematic Position

Class—Bryopsida

Order—Polytricales

Family—Polytrichaceae

Genus—Polytrichum

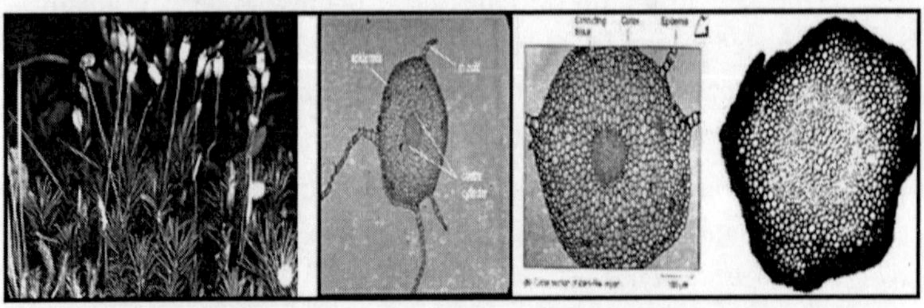

Fig. Plant Body of Polytrichum, Section of Stem

1. Commonly grows in cold and rainy season on woody logs, tree trunks, rocks and roof tops. Plant body erects and differentiates into leaf, stem and rhizoids. Rhizoids are branched filamentous with oblique walls. Stem erects wholly covered by leaves. Leaves arranged spirally on the axis, apical tufts, leaf entire, sessile, single mid-rib present. Sporophyte is present on the gametophyte.
2. A section of thallus shows, a single layered epidermis cortex composed of two zones, outer sclerenchymatous and inner parenchymatous layer. Central thick zone, has the the vascular strand.
3. Reproduction sexual, monoecious. Male sex organs formed at the apex of main shoot while the female sex organs are present at the lateral branches coming out of the main branch.

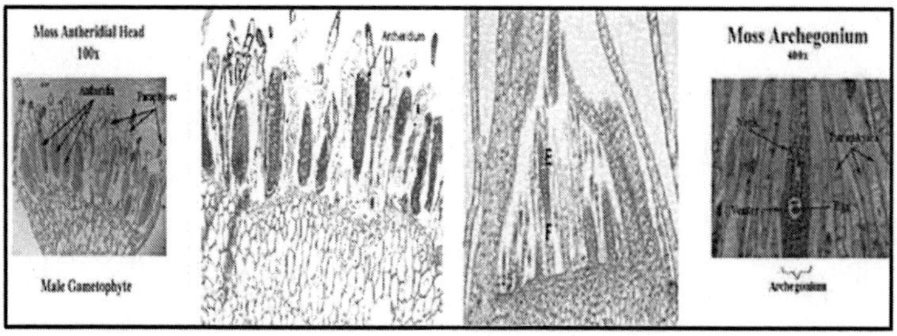

Fig. Section of Gametophyte of Polytrichum Showing Sex Organs

3. A convex disc is present at the apex forming a rossette, the apical part is occupied by antheridia and paraphyses. Each antheridium is club shaped, stalk and body present. Numerous androcytes are enclosed, paraphyses globose, multicellular. Archegonia, composed of venter canal cell, an egg, along neck with neck canal cells. Venter is surrounded by two-layered jacket. Paraphyses, simple, multicellular and filamentous.

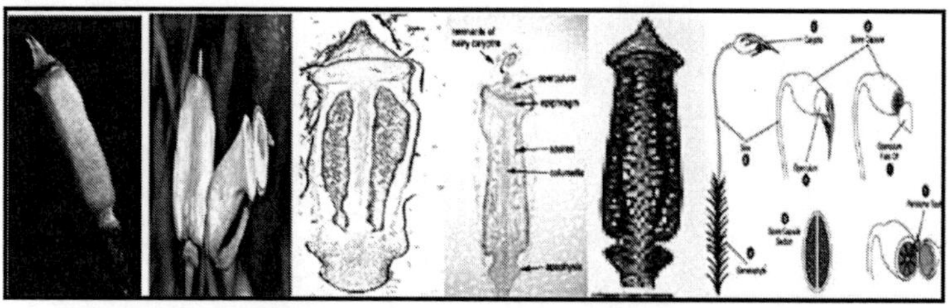

Fig. Section of Sporophyte of Polytrichum Showing Spores

5. Sporophyte consists of foot, seta and capsule.The foot remains embedded at the apex of the archegonium. Seta composed of epidermis, cortex and central tissue. Capsule, pear-shaped lower part the apophysis, sterile tissue, continues in the upper region as columella.This columella is enclosed by a spore sac guarded by wall layers. Sporogenous cells are present in spore sac, forms spore after division. An opening called as operculum is present, arc shaped, characteristic feature. A constriction separates operculum from capsule. peristome is situated, below the operculum two rows of sixteen teeth each constitutes the peristome.

VII PTERIDOPHYTA

(I) Lycopodineae

(a) Lycopodium

Systematic Position

Class—Lycopsida

Order—Lycopodiales

Family—Lycopodiaceae

Genus—Lycopodium

Fig. Sporophyte of Lycopodium Showing Strobilus

1. Lycopodium meaning- lukos, wolf and podion, diminutive of pous, foot is a genus of clubmosses, is also known as ground pines or creeping cedar. Found in tropical regions, some were in temperate regions also. Sporophyte flowerless, vascular, terrestrial, small herbaceous, shrubby. Stem prostrate with upright branches, with small, simple, needle-like or scale-like leaves that cover the stem and branches thickly; tips with strobilus. Microphyllous leaves arranged spirally or in whorls, lanceolate in shape. Some species are epiphytic in nature.
2. The kidney-shaped or reniform spore-cases, sporangia contain spores of one kind only isosporous, homosporous and are borne on the upper surface of the leaf blade of specialised leaves sporophylls arranged in a cone-like strobilus at the end of upright stems. The club-shaped appearance of these fertile stems gives the club mosses their common name.

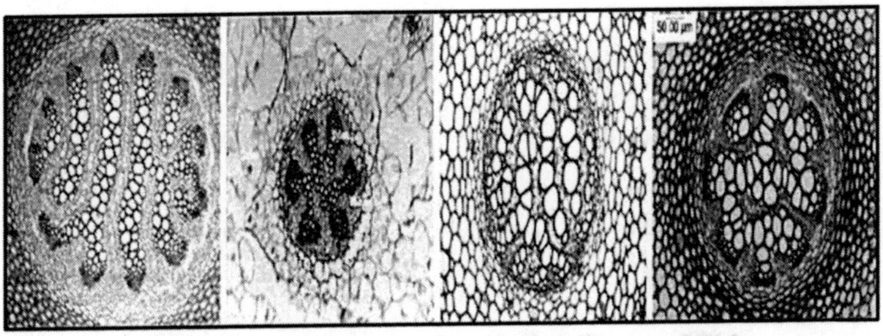

Fig. Section of Stem Showing Various Types of Steles

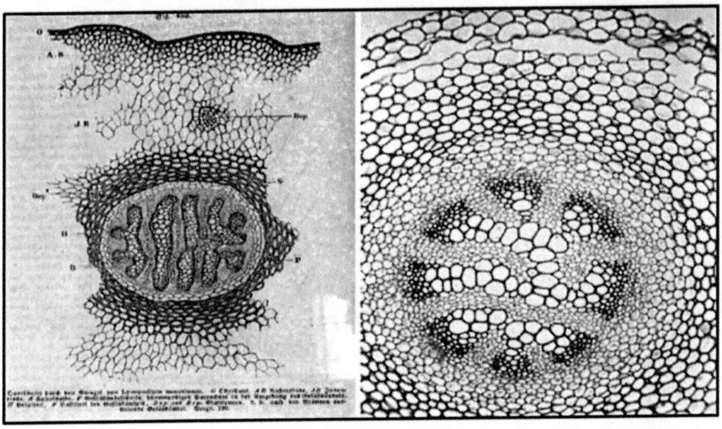

Fig. Enlarged Sector of Stem Showing Plecto and Actino Stele

3. Stele or vascular elements were found for first time in pteridophytes, called as vascular cryptogams. Xylem and phloem are surrounded by endodermis and pericycle with a central pith forming the vascular bundle. A range of stele are seen in lycopodium, haplostele, actinostele, plectostele, mixed protostele. The function of xylem being conduction of water and phloem, is as food materials.

4. A section of stem shows an outermost epidermis, interrupted by stomata, cortex variable is made of homogenous or heterogenous cells. The ensuing layer is endodermis with thickened radial walls. Pericycle is three to six cells in thickness. There is a great variation in the organisation of the vascular elements. Small core of xylem is surrounded by phloem, protostele in some xylem has radiating ribs or present in the form of plates, parallel to each other, plectostele, star shaped xylem surrounded by phloem, actinostele.

5. Lycopods reproduce asexually by spores. The plant has an underground sexual phase that produces gametes, gametophyte and this alternates in the lifecycle with the spore-producing plant, sporophyte. The prothallium or gametophyte developed from the spore is a subterranean mass of tissue of considerable size and bears both the male and female organs antheridia and archegoniae. However, vegetative reproduction is by underground rhizomes.

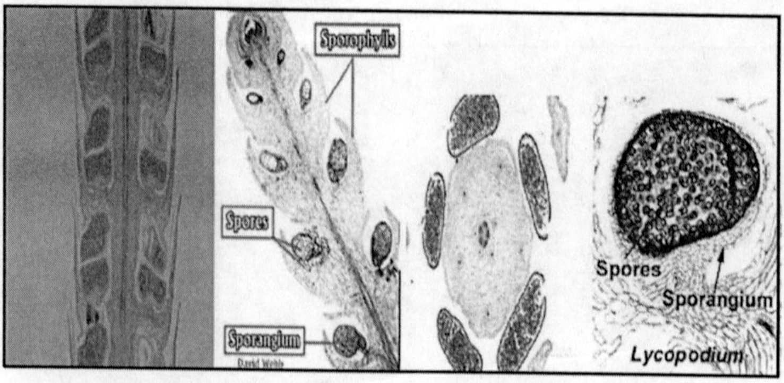

Fig. Strobilus of Lycopodium Showing Homospores

6. A strobilus is a structure present on many land plant species consisting of sporangia-bearing structures densely aggregated along a stem. Strobili are often called cones. Strobili are characterised by a central axis, anatomically a stem is surrounded by spirally arranged or decussate structures that may be modified leaves that bear sporangia are called sporophylls, while sporangia-bearing stems are called sporangiophores. Spores of uniform shape and size homosporous.

(b) Selaginella

Systematic Position

Class—Lycopsida

Order—Selaginellales

Family—Selaginellaceae

Genus—Selaginella

1. Sporophyte herbaceous is perennial or annual, found in tropical regions, shady habits. Some grow on damp shady hills in temperate regions. Under dry conditions, some species of Selaginella roll into brown balls-a phenomenon known as poikilohydry. Under moist conditions, the brown balls become green, because of which these are also known as resurrection plants.

Fig. Sporophyte of Selaginella, Habit

2. Selaginella species are creeping or ascendant plants with simple, scale-like leaves microphylls on branching stems from which roots arise. The plants are heterosporous (megaspores and microspores), and have structures called ligules, scale-like outgrowths near the base of the upper surface of each sporophyll. Rhizophores are colourless, leafless cylindrical appendages borne on the shoot.

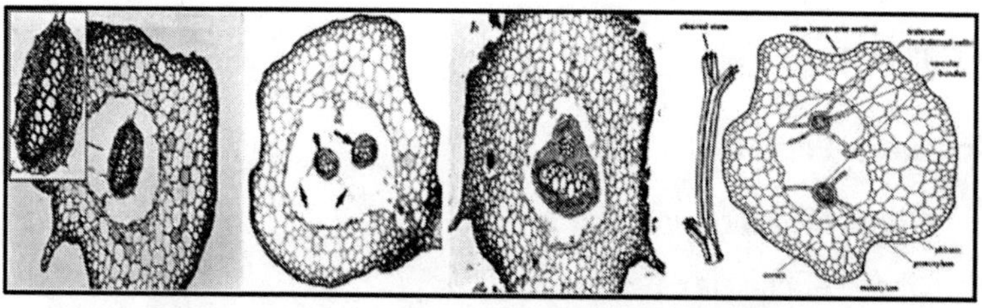

Fig. Transverse Section of Stem of Selaginella

3. A section of stem shows outer layer epidermis, prosenchymatous and cutinised. This is followed by a hypodermis, thick walled cells next is the cortex with-thin walled parenchymatous cells. Endodermis trabeculated stretched cells with intercellular spaces. Pericycle is hin-walled. Stele is protostele, polystelic. Xylem is band shaped with elongated cells have tracheids only, exarch. Xylem is surrounded by phloem.

4. The cone or strobilus has a central axis on which is pointed sporophylls are borne. Axil of each sporophylla is present sporangium Outer wall is thick, inner wall is thin. Sporangia are of two types-microsporangium and megasporangium.

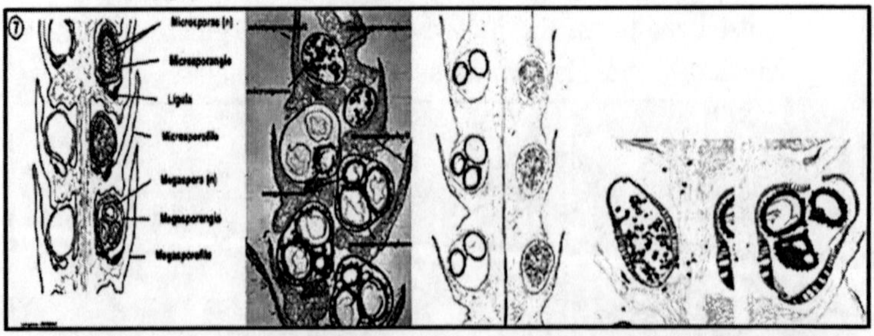

Fig. Section of Strobilus of Selaginella Showing Heterospores

5. Microsporangium are borne on terminal portion of strobilus. It produces number of small-sized microspores, arranged in tetrads. Each microspore has an exine and in intine. Megasporangium are present towards the base they are four in number, large in size, exert pressure on the walls to give a four-lobed shape to the sporangium.

(II) Equisetineae

(a) Equisetum

Systematic Position

Class—Sphenopsida

Order—Equisetales

Family—Equisetaceae

Genus—Equisetum

1. The name "horsetail", often used for the entire group, arose because the branched species somewhat resemble a horse's tail. Similarly, the scientific name Equisetum derives from the Latinequus "horse" and seta "bristle". Stems are coated with abrasive silicates, making them useful for scouring (cleaning) metal items such as

cooking pots or drinking mugs, particularly those made of tin. Many species in this genus prefer wet sandy soils, though some are semi-aquatic and others are adapted to wet clay soils.

Fig. Habit of Equisetum at the Nodal and Internodal Region

2. The plant body has a creeping rhizome. Aerial branches are of two types, the sterile branches are green and have whorl of lateral branches at each node, fertile branches are colourless, unbranched and have strobilus at the apex. Stem is made up of nodes and inter-nodes, alternating with each other. At each node, a whorl of scale like leaves form a sheath. Leaves small, simple, uninerved slender, teeth like, devoid of chlorophyll.

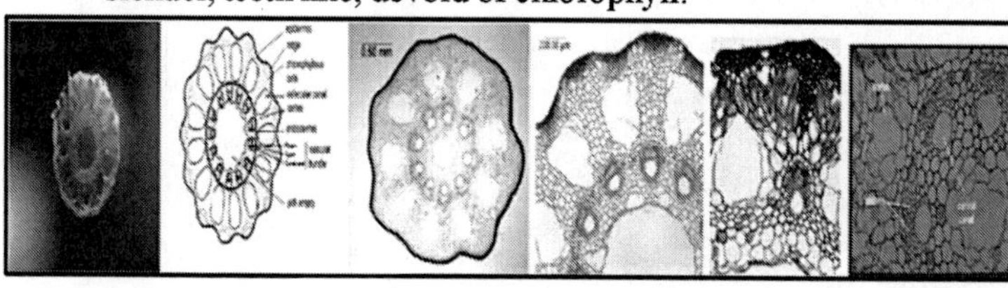

Fig. Section of Stem of Equisetum

3. Internally, the stem is differentiated into epidermis, cortex and stele. Prominent ridges and furrows are seen. Presence of stomata in furrows alone. Epidermis is followed by sclerenchymatous hypodermis, chlorenchymatous cortex under the furrow region. Inner cortex uniform, parenchymatous. Below each furrow is a vallecular canal. Stele ectopholic siphonostele. Vascular bundles are below the ridges only. Metaxylem in two concave plates protoxylem occurs on either sides of carinal canal, phloem occurs inbetween metaxylem. The central part is occupied by a pith cavity or central canal.
4. Strobilus develops on the apex of a fertile branch called as sporangiophores. Strobili have several whorls of densely crowded

peltate appendages, which bear the sporangia. A ring like structure, annulus is borne by the central axis below the sporangiophores. The sporangia are borne in a ring and on the sides of the disc facing the central axis of the strobilus. Each sporangium is elongate with the rounded apex. Sporangium bears many spores.

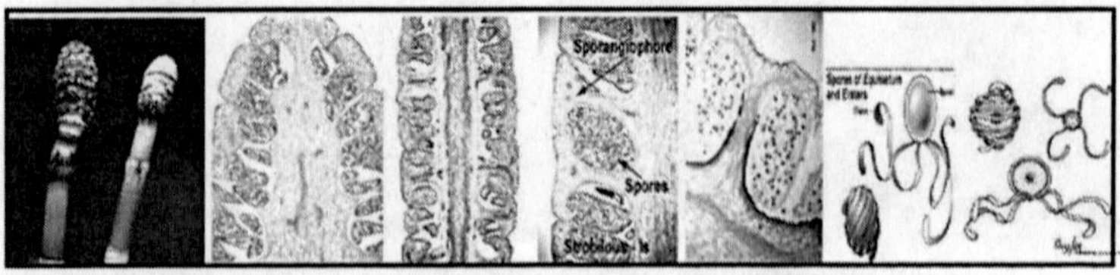

Fig. Strobilus of Equisetum Showing Homospores

5. Spores homosporous, outer layer exosporium, forms four strips are coiled about the spore until it dehisces. Elaters, four are present around each spore, helps in dispersal, hygroscopic. Spores are rich in chloroplast.

(III) Filicineae

(a) Gleichenia/Dicranopteris/Stromatopteris

Systematic Position

Class—Filicineae

Order—Filicales

Family—Dicranopteriaceae

Genus—Dicranopteris

1. A common plant, often seen growing under waterfalls, in swamps, under cliffs and in tall open forest. It prefers high humidity and good levels of sunshine and moisture Rhizome light is brown, scabrous, covered with many chestnut-coloured hairs, falling with time, primary lateral branches are formed in several opposite pairs, frond bifurcate to reniform-lunate in outline, with 1 (rarely 2) level of false dichotomy in each lateral branch system arising from each side of the terminal bud with stipules between bifurcations, pinnas variable, glaucous and glabrous on underside, occasionally

with stellate hairs on this side, veins bifurcating, sori often numerous partially immersed in the lamina, consisting of 2-4 sporangia, each in a separate but adjoining pit.

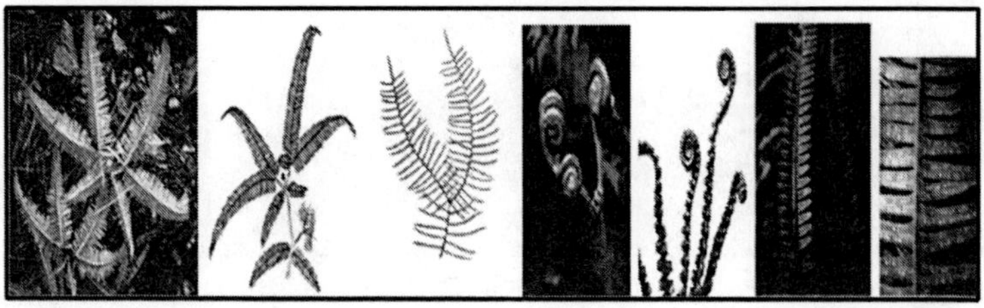

Fig. Sporophyte of Dicranopteris Showing Circinate Vernation, Sori

2. The stem of the fern plant, may creep along or under the ground or even up a tree, it may grow into a short or tall trunk a vertical rhizome or it may be a solid mass that gives rise to a tuft of fronds an erect rhizome, which grows underground.

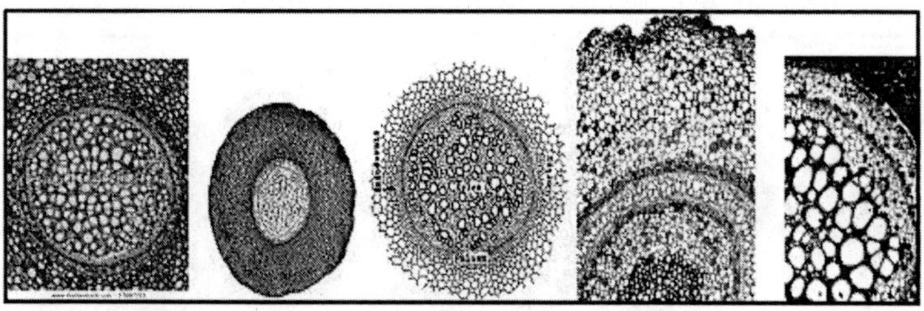

Fig. Section of Rhizome of Dicranopteris

The rhizome produces roots and new fronds. A leaf gap is where the vascular tissue branches off to a frond. In cross-section, the vascular tissue does not form a complete cylinder, where a leaf gap occurs. Epidermis is single layered, cortex heterogenous with tannin and cell deposits. Vasculature is amphipholeic siphonostele. A single ring of xylem is surrounded on both sides by phloem, an outer and inner phloem. Central pith. Xylem with vessels having perforate walls and group of sclerenchymatous fibres are well developed above the xylem.

3. The petiolar bundle in Gleichenia is primitively C-shaped and consists of an abaxial and two adaxial loops. At the abaxial end of each loop, generally there is a differentiated group of narrow vessels, which may be called an isthmus. In some leaf-axes, a false

gap at the line of junction of isthmus and abaxial are may extend for some distance, but it is independent of the exit of branch traces.

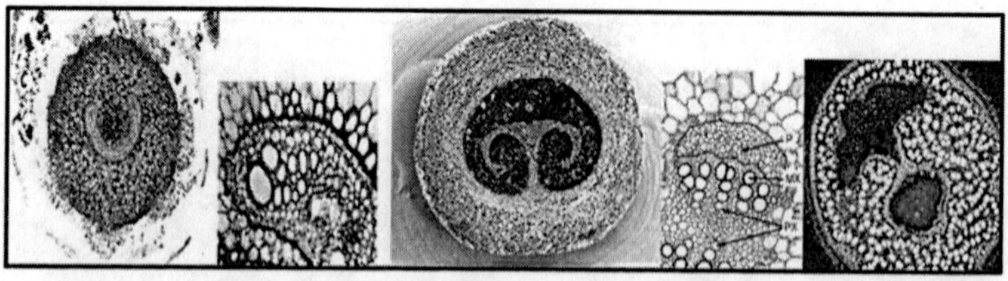

Fig. Section of Petiole of Dicranopteris

The primitive number of protoxylem groups is three, but this number is seen in certain regions of the leaf only due in part to a strong tendency for the median group to split into two in preparation for forking of the leaf-axis. The leaf-trace is typically nearly as wide as the stele at the region of insertion. All evidence indicates that the characteristic forks are not true dichotomies.

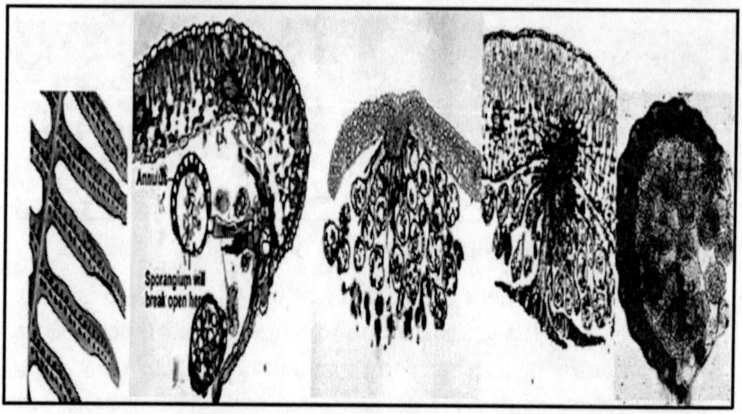

Fig. Section of Pinna of Dicranopteris Showing Sori

4. The reproductive structures are sori, found on underside of the frond. Each sporangium is a capsule that contains spores. They are usually aggregated into clusters called sori. Most ferns produce 64 spores in each sporangium. Outer wall thick and ornamented inner wall smooth and thin. When mature, the spores are released from the sporangia. Once released, the spores germinate readily on contact with damp soil.

5. Gametophyte subterranean, cylindric, mycotrophic bears rhizoids and superficial gametangia. Antheridia large, many celled produces numerous spermatozoids. Archegonia large, long necked straight or curved.

(b) Marsilea

Systematic Position

Class—Filicineae

Order—Filicales

Family—Marsileaceae

Genus—Marsilea

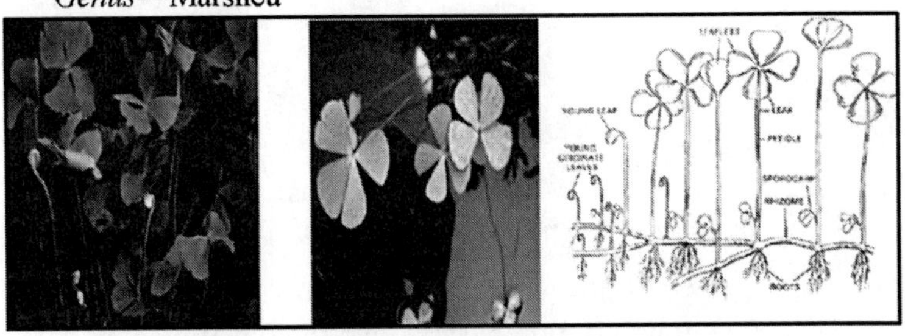

Fig. Section of Sporophyte of Marsilea

1. It is a heterosporus fern, aquatic or semi aquatic in habitat. Commonly, it is called as water clover and four-leaf clover because the long-stalked leaves have four clover-like lobes, either held above water or submerged. Have long, slender rhizomes that creep beneath the ground. The plants appear to have more stem support than leaf, unlike other ferns. Vessels have also been found in the rhizome. The four leaflets on the leaf are not borne equally. Instead, they are borne in pairs with one pair of leaflets attached slightly higher than the other. Thus in the developing leaf, the leaflets are folded more like the wings of a butterfly than like the leaflets of a clover.
2. In addition to their vegetative sterile leaflets, it produces fertile spore-producing leaflets at or near the base of the photosynthetic leaves. This reproductive portion looks and functions very differently from the vegetative portion of the leaves. Wall is nothing but a modified blade of pinna. This structure is called as

sporocarp, it produces both micro and macrosporangia. These are hairy, short-stalked bean-shaped structures usually 3 to 8 mm in diameter with a hardened outer covering. This outer covering is tough and resistant to drying out, allowing the spores inside to survive unfavourable conditions such as winter frost or summer desiccation. Despite this toughness, the sporocarps will open readily in water, if conditions are favourable, and specimens have been successfully germinated after being stored for more than 130 years.

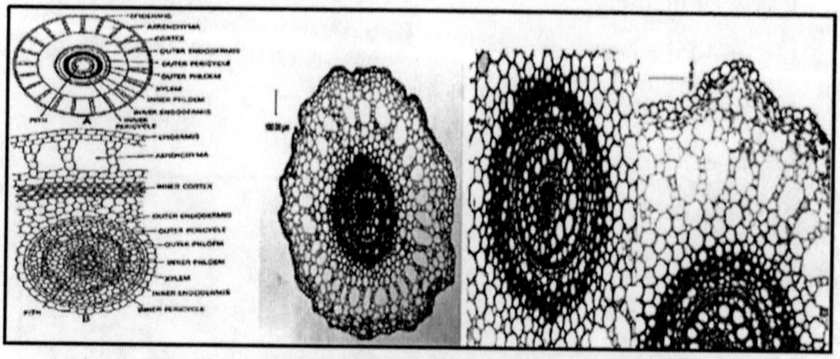

Fig. Transverse Section of Rhizome of Marsilea

3. A section of rhizome shows outermost epidermis, cortex heterogenous, outer cortex, parenchymatous, middle sclerenchymatous, inner compact parenchymatous. In between the middle and outer cortex, there lies air chambers separated by trabeculae. Stele amphiphloic solenostele have phloem on both sides of xylem. Next to inner phloem is the inner pericycle and inner endodermis and central pith.

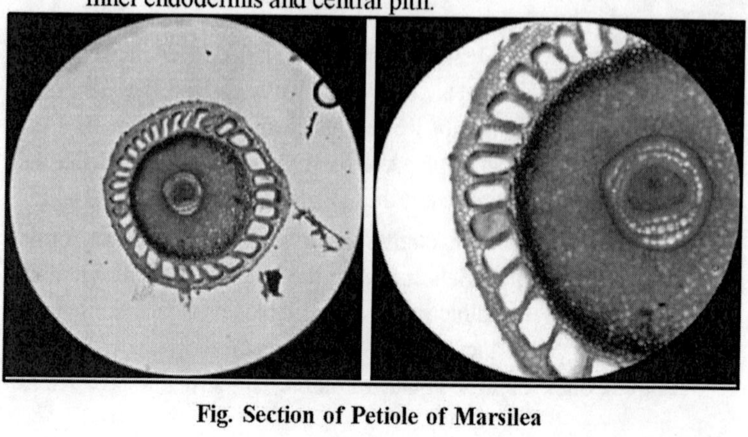

Fig. Section of Petiole of Marsilea

4. Petiole of Marsilea shows one cell epidermal layer, hypodermis parenchymatous with conspicuous air spaces septa or trabeculae partition air space into air chamber. Cortex, outer sclerenchymatous inner parenchymatous, stele is bounded by monolayer of endodermis, pericycle, xylem with two arms metaxylem in the centre protoxylem at either ends. Xylem, surrounded by phloem is protostelic in nature, but pith absent.

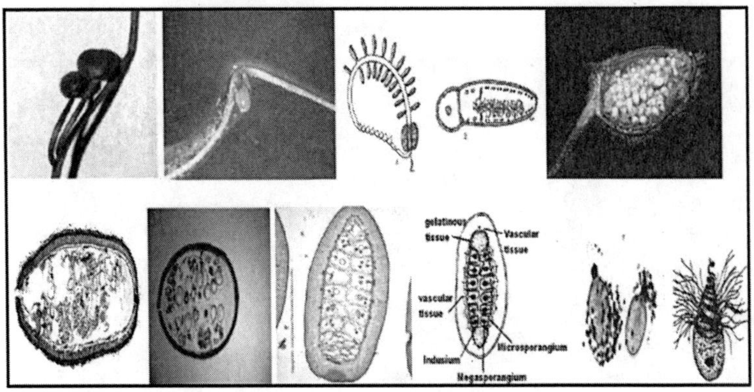

Fig. Section of Sporocarp of Marsilea Showing Heterospores

5. Vegetative reproduction is by tubers. Asexual by spores. Each sorus includes a mix of two types of sporangium, each type producing only one of two kinds of spores. Towards the center of each sorus and developing first are the megasporangia, each of which will produce a single large female megaspore. Surrounding them at the edge of the sorus and developing later are the microsporangia, each of which will produce many small male microspores. It produces two kinds of spore and thus have two kinds of gametophyte, they are called heterosporous.

Dehiscence of sporocarp is special. It splits into halves, allowing the tissue coiled inside to become hydrated, tissue swells with water. It pushes the halves of the hard outer covering apart, and emerges as a long gelatinous worm-like sorophore. The sorophore is a sorus-bearing structure unique, it may extend to more than ten times the length of the sporocarp thin which it was coiled. This extension carries the numerous spore-producing sori attached along each side of the sorophore out into the water.

(c) Azolla

Systematic Position

Class—Pteridopsida

Order—Salviniales

Family—Salviniaceae

Genus—Azolla

Fig. Section of Sporophyte of Azolla

1. It is a water fern of stagnant water, reduced in form and specialised to fix atmospheric nitrogen. Consists of a main rhizome, which branches into secondary rhizomes which bears alternately arranged small leaves. Ventrally, unbranched adventitious roots hang down into the water from nodes. Nutrients are absorbed directly from the water by the roots. It can spread rapidly and has the ability to survive on moist soil in and around rivers, ditches and ponds. In eutrophic waters, it forms dense surface mats, which degrade water quality by reducing oxygen levels.
2. The plants are small and float, but can be very abundant and form large mats. The plants are typically red, and have very small water repellent leaves. Azolla floats on the surface of water by means of numerous, small, closely overlapping scale-like leaves, form a symbiotic relationship with the cyanobacterium Anabaena azollae, which fixes atmospheric nitrogen, giving the plant access to the essential nutrient. This has led to the plant being dubbed a "super-plant", as it can readily colonise areas of freshwater, and

grow at great speed—doubling its biomass every two to three days.

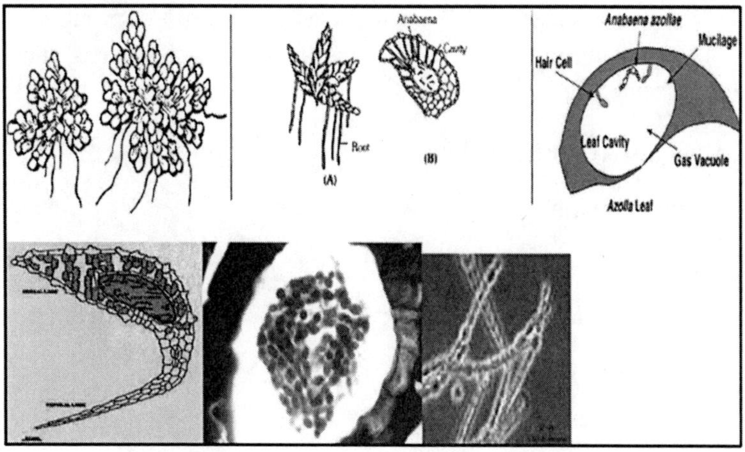

Fig. Frond Section of Azolla Showing Endophytic Cyanobacteria

3. The blue-green alga Anabaena azollae consists of unbranched trichomes containing bead-like, heavily pigmented vegetative cells and lightly pigmented, intercalary heterocysts, which are slightly larger and have thicker cell walls. In very young leaves, trichomes lack heterocysts. Heterocysts gradually increase in frequency until they comprise 30-40 per cent of the algal cells. Mature trichomes also contain spores called akinetes.
4. Vegetative and asexual reproduction is absent. During the summer months, numerous spherical structures called sporocarps form on the undersides of the branches. The male sporocarp is greenish or reddish and looks like the egg mass of an insect or spider with numerous male sporangia. Male spores microspores are extremely small and are produced inside each microsporangium, tend to adhere in clumps called massulae. Female sporocarps are much smaller, containing one sporangium and one functional spore. Since an individual female spore is considerably larger than a male spore, it is termed a megaspore.
5. Azolla has microscopic male and female gametophytes that develop inside the male and female spores. The female gametophyte protrudes from the megaspore and bears a small number of

archegonia, each containing a single egg. The microspore forms a male gametophyte with single antheridium, which produces eight swimming sperm. The barbed glochidia on the male spore clusters are assumed to cause them cling the female megaspores, thus facilitating the fertilisation.

VIII GYMNOSPERMAE

(A) Cycadopsida

(i) Cycas

Systematic Position

Class—Cycadopsida

Order—Cycadales

Family—Cycadaceae

Genus—Cycas

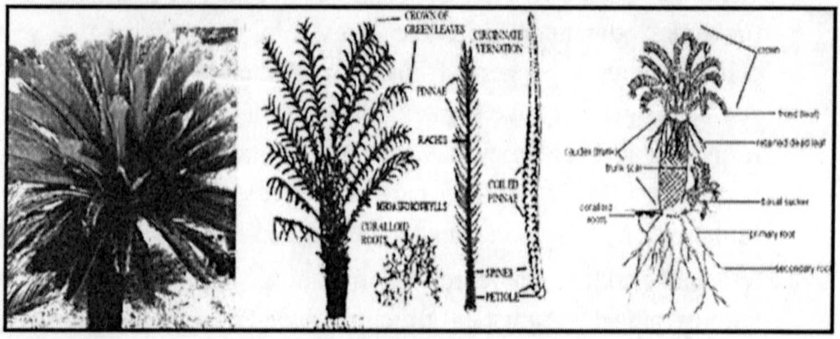

Fig. Habit of Cycas

1. The plant belongs to the sporophytic generation and is differentiated into root, stem and leaves. The adult plant is about two meters in Right. The stem is thick, unbranched, caudex, erect, stout, but branches arise when terminal bud is injured and bears triangular leaf bases on its surface. The leaves show circinate vernation and are loosely arranged. Cycas leaves are of two types (a) scale leaves (b) green foliage leaves. Scale leaves are short, dry and brown covered with hairs. Foliage leaf, pinnately compound with the expanded leaf base.

2. The roots are of two types, normal tap root with lateral rootlets and were dichotomously branched coralloid roots. The primary root persists and forms a tap root system.

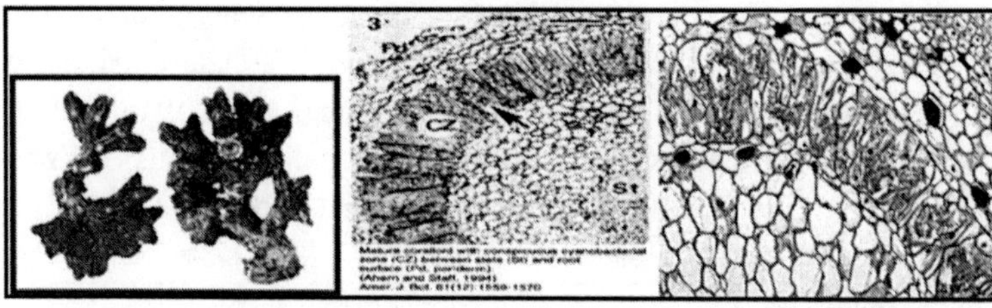

Fig. Coralloid Root: Entire, Section of Root Showing Algal Colony

The lateral branches of the root come out of the soil (apo-geotropic) and branch repeatedly to form dwarf dichotomously branched coral like man called coralloid roots or corallrhiza. Blue green algae Anabaena live symbiotically in the coralloid roots. It fixes the atmospheric nitrogen. Epiblema single layered, cortex well differentiated into outer, inner and mid cortex. The cyanobacteria are present in the mid-cortex. Endodermis and pericycle single layered, vascular bundle radial, protoxylem exarch, diarch to tetrarch. Pith may or may not be present in the from.

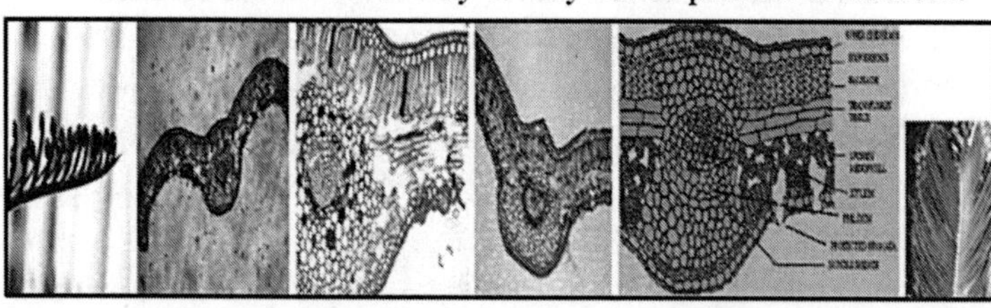

Fig. Section of Leaflet of Cycas Showing Transfusion Tissue

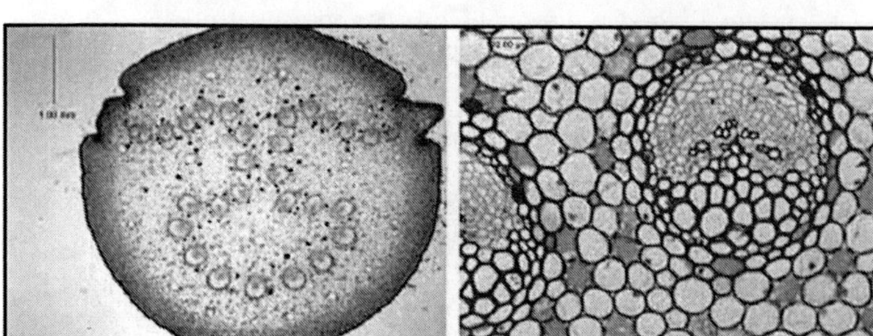

Fig. Section of Rachis of Cycas Showing Enlarged Vascular Bundle

3. The petiole is stout and called rachis. Rachis bears leaflets or pinnules on either side. The lower leaf-lets of rachis are modified into spine. Lateral veins are absent in is pinnule. Rachis shows anatomically, outer hypodermis, stomata present, ground tissue has parenchymatous with mucilage ducts. Vascular bundles are arranged in an omega shaped manner. Vasculature with sclerenchymatous have bundle sheath, conjoint, collateral and open. Xylem diploxylic, centripetal and centrifugal types are present, mesarch, phloem is present.

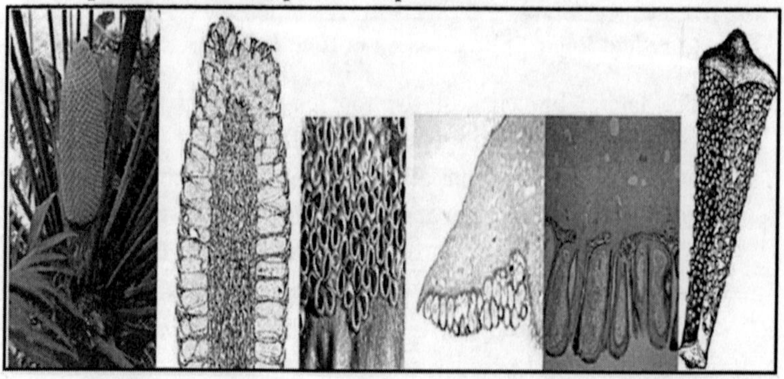

Fig. Section of Male Cone of Cycas Showing Microspores

4. Leaflet with single layer of epidermis, cuticle present, tough and leathery texture, stomata is sunken on lower epidermis. Mesophyll is well differentiated into upper palisade and lower spongy tissue. Transfusion tissue helps in conduction, cells colourless, elongated run laterally in the centre, point to a xerophytic adaption .Vascular bundle is conjoint, collateral, xylem is diploxylic, the larger patch is the centripetal xylem and the two lateral patches are centrifugal xylem. Phloem is present below the centrifugal xylem.

Fig. Female Cone of Cycas Showing Ovule

4. The vascular bundles in stem are conjoint, collateral open and arranged in a ring. The xylem of the bundles consists of tracheids only. The pholem comprises sieve tube and phloem parechyma, they lack from companion cells. The secondary growth in a young stem is normal, but as it becomes older, new rings are formed giving rise to concentric rings of secondary vascular tissues (polyxylic condition).

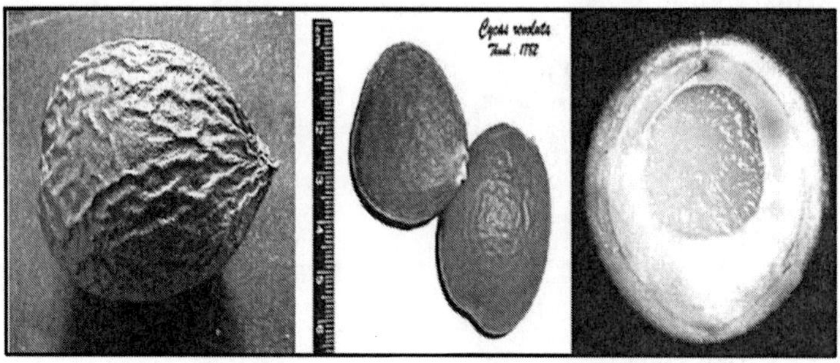

Fig. Seed of Cycas: Entire, Section

5. Cycas is dioecious. The male plant has a cone at its apex. Each cone has a thick central axis around which microsporophylls are spirally arranged. Groups of sporangia or sori are present on the abaxial (lower) side of the sporophyll. Each sorus contains two--six sporangia or anthers which are filled up with pollen grains or microspores. Microspore represents the first cell of male gametophyte stage. The upper part of microsporophyll is sterile and pointed is called apophysis.

   There is no true female cone is cycas. The megasporophyll arises at the apex of the female plant and are loosely arranged in rosette manner. On either side of the broader part of sporophyll are four—six red coloured ovules or megasporangia arranged. Cycas ovule has single, thick threelayered integument which, is hard and inner layers are fleshy and the middle layer is hard and stony. Deep in the nucellus of the mature ovules lies the megaspore mother cell, which divides into four megaspores. Only one of the four megaspores is functional and rest of the three degenerate. By repeated divisions of the functional megaspore, female gametophyte is formed. On the upper surface of the female gametophyte, two–eight archegonia are formed, each having a ventral cell and an egg. Seed hard with integuments, embryo lies in the tissue of endosperm.

(B) Coniferopsida

(i) Pinus

Systematic Position

Class—Coniferopsida

Order—Pinales

Family—Pinaceae

Genus—Pinus

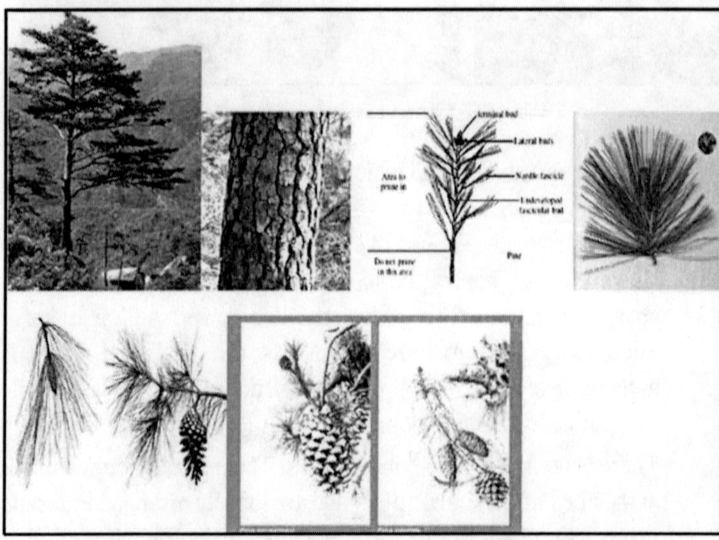

Fig. Habit of Pinus Showing Long and Dwarf Shoots

1. The Pinus tree represents the sporophytic generation, gregarious in habit. The stem displays the excurrent habitat with spiral growth of branches. The main stem is branched. Branches are of two types. Branches with unlimited growth, grow by means of apical bud. Branches with limited growth, arising directly from the trunk. Leaves are also of two kinds, Foliage leaves, are unusual type-being long, narrow, tough, green and are frequently known

as Pinus-spur or Pine-needles. They are borne only on the dwarf shoots in clusters of two, three or five. Branches with foliage leaves are called spurs. Spurs could be monofoliar, bifoliar, trifoliar, pentafoliar having one, two, three, five foliar leaves. The leave types are scale leaves. These are brown, membranous and are protective in function. These are borne on both type of branches, but fall off as the dwarf shoots mature. Scale leaves of dwarf shoots are called cataphylls. Primary root persists and forms a typical elongated straight tap root, Mycorrhizal association seen. Monoecious i.e., male and female cones are found in the same plant, reproduces by asexual spores, which are produced in sporangia borne on specialised structures, the sporophylls.

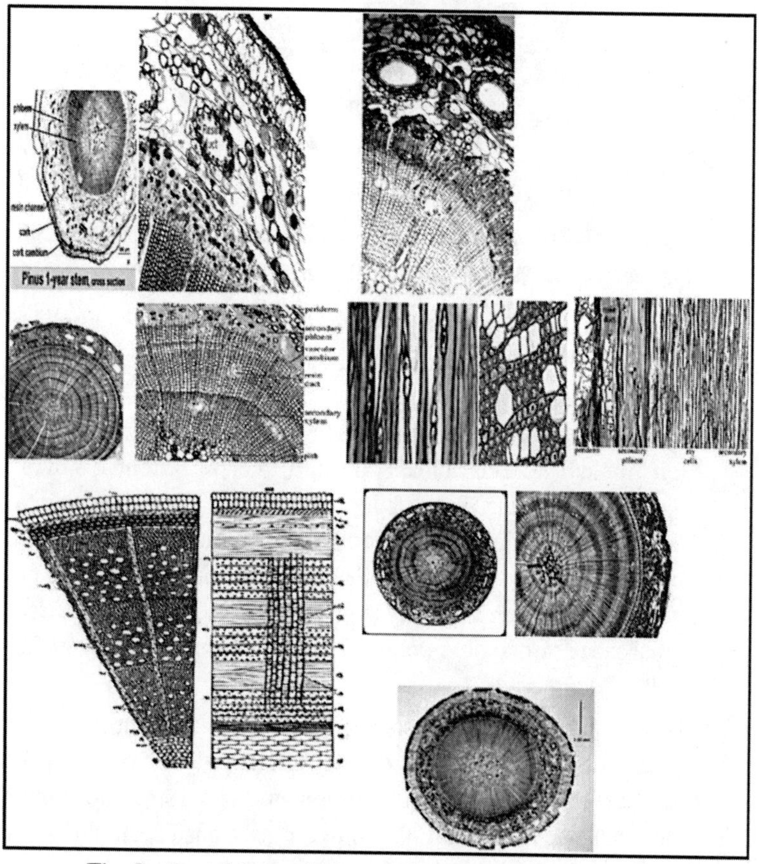

Fig. Section of Stem of Pinus Showing Secondary Growth

cuticle thick, cortex parenchymatous with intercellular spaces. Resin ducts common in cortex. Vascular bundles open, collateral and endarch. Medullary rays narrow and pith parenchymatous. Secondary growth seen, wood pycnoxylic. Cork outermost layer, cork cambium two to three cell in thickness, followed by secondary and primary cortex. Primary phloem and secondary is not differentiated. Cambium is well marked, secondary xylem with tracheids, small summer wood and large spring wood. One spring and summer wood constitutes an annual ring. Medullary rays are prominent. Secondary xylem with many resin canals are surrounded by an epithelium. Pith is present, with traces of primary xylem. Medullary rays are of two kinds normal rays and fusiform rays.

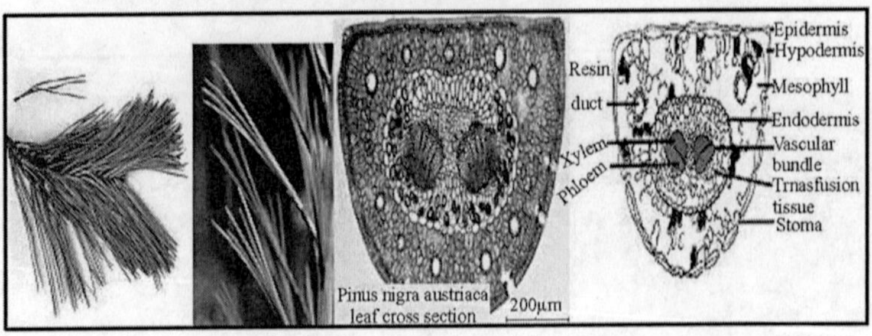

Fig. Section of Needle of Pinus Showing Transfusion Tissue

3. Needle is triangular in shape. Outermost epidermis is thickwalled with many sunken stomata. Hypodermis is sclerenchymatous, resin is canal seen, each resin canal doublewalled, hypodermis is followed by mesophyll, parenchymatous cells are rich in chlorophyll, vascular bundle are in the centre surrounded by a layer of endodermis, has multilayered pericycle. Bundles conjoint, are closed, two due to formation of transfusion tissue in between two bundles. Conduction is the main function of transfusion tissue, lack of veins. Primary xylem contains neither true vessels, tracheae nor wood fibres, only tracheids phloem lacks companion cell.

4. Male cone occurs in clusters near the ends of long branches axil of short branches, strobilus develops in the axil of the bract and replaces the spur. Each male cone is simple ovoid structure-about one inch long with 60 to 100 spirally arranged microsporophylls or

stamens. They arise from the central axis and pass out horizontally. Each microsporophyll stamen has a short stalk and much longer expanded scale like part on the under surface of horizontal part are borne two microsporangia or pollen sacs.

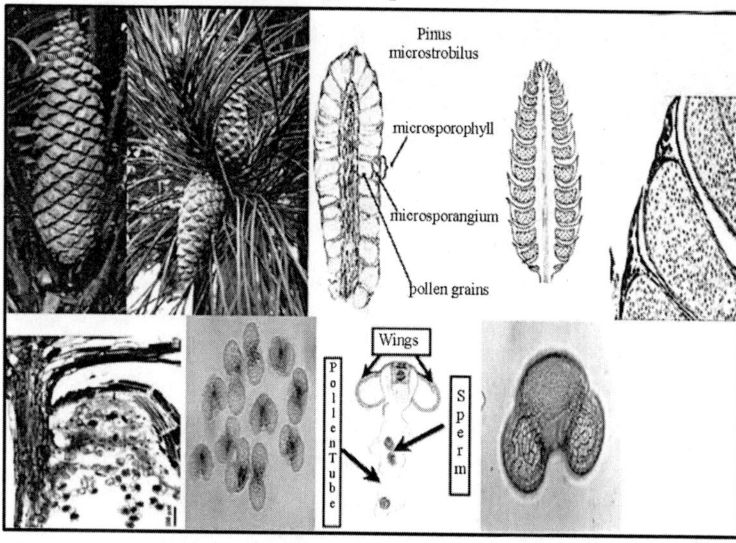

Fig. Section of Male Cone Showing Microspores, Winged Pollen Grains

Inside the sporangia, large number of microspore mother cells is present, which undergo reduction division and produce haploid microspores or pollen grains. Each pollen has two wings on the lateral sides.

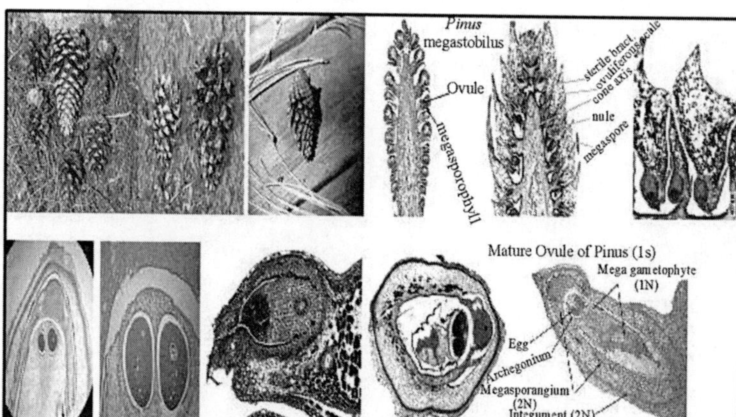

Fig. Section of Female Cone Showing Ovule

5. Female cones are simple, bigger, woody, dry and hard structures, finds developed laterally in the axils of scale leaves.

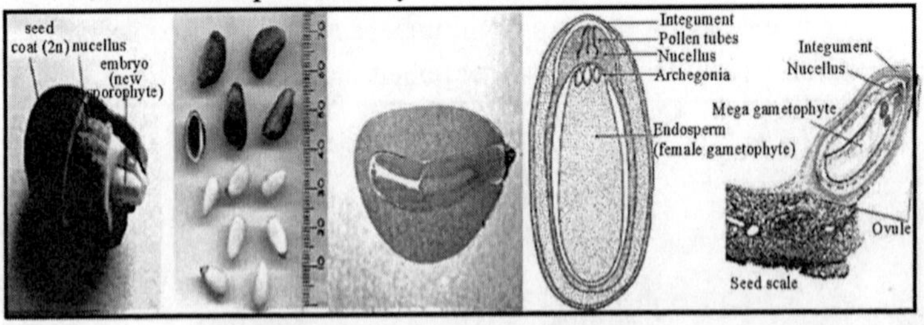

Fig. Seed of Pinus: Entire, Section

They are usually produced in clusters in place of shoots of unlimited growth and on different branches from the male cones. Each female cone has central axis to which are attached megasporophylls. Each meagsporphyll on its upper surface have two ovules. Each megasporophyll has an outer small bract scale and a large ovulifervous scale. Bract scale arises directly from the bract scale. The ovule arises as a group of cells, which form rounded hump of tissue termed is the nucellus megasporganum. From the base of nucellus, an endosing integument grows up which does not enclose nucellus completely, but leaves an opening the micropyle which leads to the outer end of nucellus. Ovules anatropous, two Archegonia with a small neck, venter canal cell and an egg. Spores form the gametophytic phase of life cycle. In Pinus spores are of two types microspores and megaspores, which on germination give rise to male and female gametophyte. Seeds develop on ovuliferous scales, endospermic and winged.

(C) Gnetopsida

(i) Gnetum

Systematic Position

Class—Gnetopsida

Order—Gnetales

Family—Gnetaceae

Genus—Gnetum

1. It is a small to medium-size tree some are lianas. The leaves are evergreen, opposite, long and broad, entire, emerging bronze coloured, maturing glossy dark green.

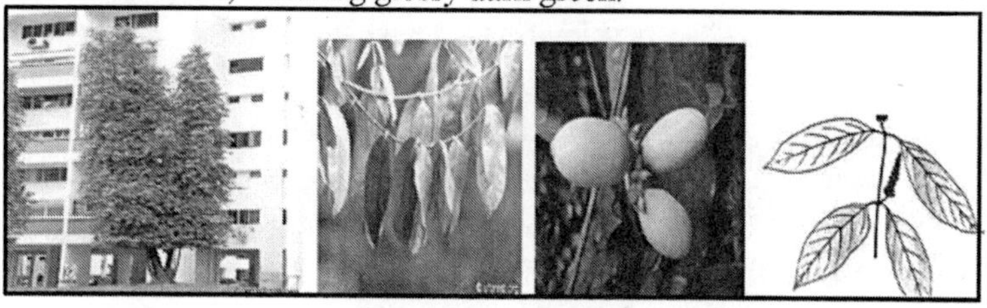

Fig. Gnetum Habit

Mature leaves appear simple; broad and flat elliptic, angiosperm-like, with reticulate venation, exstipulate, leathery, with hydathodes, not clustered, opposite and decussate. Monoecious or dioecious mostly. The reproductive structures are organised like reduced angiosperm flowers with stamen-like antherophores, but with exposed, gymnospermous ovules. The female flowers borne in spikes, which are often grouped into more complex inflorescences. The spikes bear decussate bracts, in the axils of which are condensed partial inflorescences of about three-eight female flowers, which form whorls round the stem and are intermingled with numerous hairlike structures. The fruit-like strobilus consist of little but skiny and a large nut-like seed long inside. Strobili mature mainly from the mpnth of June, every you.

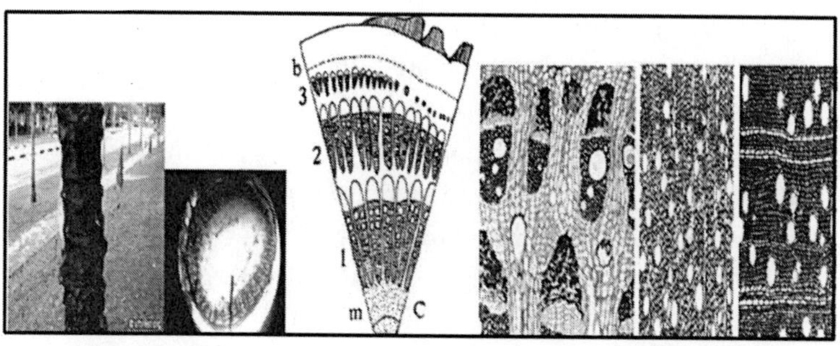

Fig. Section of Stem Showing Anomalous Secondary Growth

2. Young stem with conjoint, collateral, open and endarch bundles. In the stem the, general arrangement of various tissues and its pattern of secondary growth are very similar to those of angiosperms. The

most conspicuous similarity lies in that the xylem contains vessels and the phloem, sieve elements and companion ceils. In climbing species, secondary growth initiates in normal manner which is followed by the development of new cambium at various loci among the parenchyma cells towards the periphery of each bundle. Gradually, these loci become incorporated into a continuous cylinder, producing a normally oriented ring of xylem and phloem separated by broad medullary rays. The growth of the first ring ceases at the commencement of the further formation of the outer, successive rings. Sometimes, the ring of vascular bundle is incomplete, resulting in the formation of accentric rings.

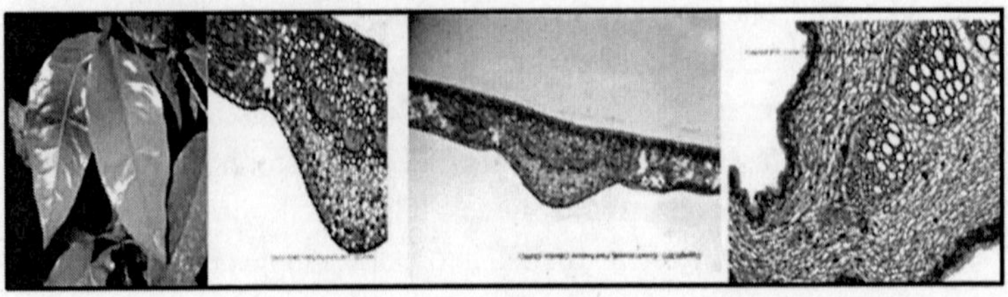

Fig. Section of Leaf Showing Xylem Vessels

3. Leaf dorsi ventral divisible into three regions epidermis, mesophyll and vascular bundle. Epidermis single layered cells are undulating, stomata on lower epidermis only. Mesophyll into palisade tissue and spongy tissue, are rich in chloroplast. The vascular bundle in the mid vein consists of an arc of bundles, conjoint, collateral, endarch, xylem towards upper epidermis. Vascular bundle is enclosed by a ring of sclerenchyma. Venation reticulates with a mid-rib.
4. Reproductive organs known as cones or strobili arise in the axil of paired scale leaves, fuse together to form a cup like structure, bract. Cluster of sporangia specialised in structure hence are called as inflorescence. The male flowers in spikes, inflorescences. The spike bears decussate bracts in the axils of which are condensed, partial inflorescences of up to 40 flowers. The effectively whorled flowers are intermingled with hair like structures. At the top of each nodal group in the male inflorescence there is usually single ring of female flowers, which are rarely fertile. Male flowers with a tubular two-membered perianth, and at

the top of the projecting axis, right and left, are two sessile unilocular anthers. Pollen are without air bladders.

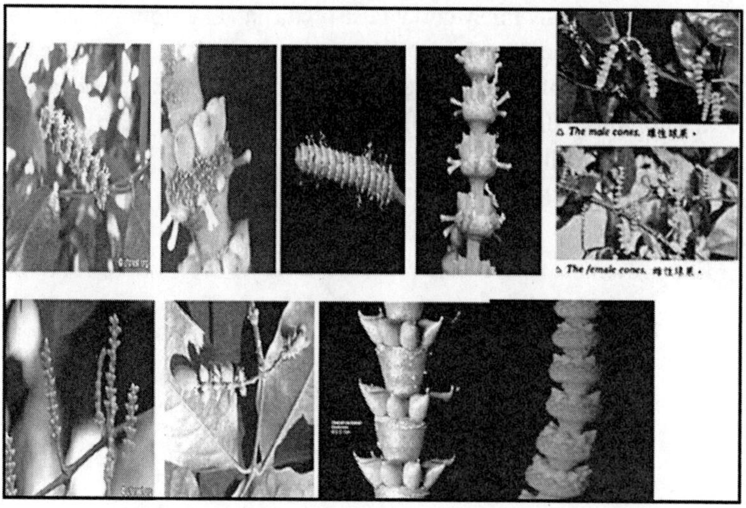

Fig. Male and Female Inflorescence of Gnetum

Pollination is mostly anemophilous, but entomophily associated with nectar secretion and is known in a few species, involving a "liquid drop" mechanism. The female flowers are borne in spikes, bear decussate bracts in the axils of which are condensed partial inflorescences of about three-eight female flowers, which form whorls round the stem and are intermingled with numerous hairlike structures. Ovule borne erect orthotropus with a long, projecting, tubular micropyle, two integumented, if the outer layer is interpreted as a perianth, or three integumented.

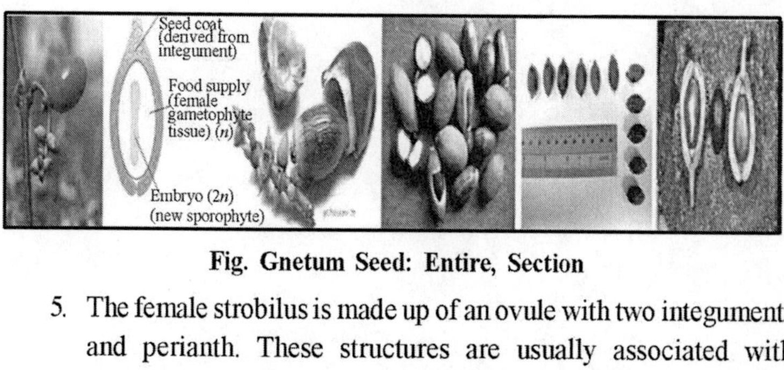

Fig. Gnetum Seed: Entire, Section

5. The female strobilus is made up of an ovule with two integuments and perianth. These structures are usually associated with angiosperms, one of the points that traditionally places Gnetum in

an ambiguous state intermediate between the gymnosperms and angiosperms. Seeds drupe like, enclosed in a red, orange, or yellow, fleshy rarely corky false seed coat, after fertilisation, the perianth becomes fleshy, the outer integument becomes woody, and the whole forms a drupe-like fruit, female gametophyte tissue copious, succulent, polyembryony seen. Cotyledons two. Germination epigeal.

Glossary of Botanical Terms

A

Abaxial: The side away from the axis, for instance the lower surface of a flowering plant leaf.

Achlorophyllous: devoid of chlorophyll.

Acervulus: cushion like mass of hyphae, forming conidiophores, conidia at the tip.

Adaxial: The side next to the axis; e.g. the upper surface of a flowering plant leaf.

Aecium: cup shaped sorus of rust forming aeciospores.

Acroscopic: facing or directed towards the apex of the axis on which the structure is borne.

Acrostichoid: of sori, producing sporangia apparently or actually spread across the surface of the fertile lamina.

Actinostele: a radially arranged, lobed protostele associated with microphylls, commonly found in the Lycopodiaceae.

Aerophore: a spot, swelling, or fingerlike projection of thin-walled cells found along the stipe or in the pinna or pinnule axils of some ferns, especially Thelypteridaceae; more prominent on young fronds.

Akinete: non-motile reproductive cells with thick walls.

Alate: of spores, with the outer wall exospore or perispore is raised in a pattern of narrow, winglike ridges surrounding depressions.

Alete: of spores, is spherical, lacking from a laesura, and containing more sets of chromosomes than normal spores, found regularly or occasionally in some apomictic ferns.

Anadromous: with the basal pinnule and/or vein group of the pinna is directed toward the frond apex.

Anisophyllous: bearing fronds or microphylls of unequal size, but is of a single shape.

Annulus: a row or patch of partially or entirely thick-walled cells of the capsule of the leptosporangium, which contracts and forces the capsule to open and to discharge its spores.

Antheridium: adjective antheridial- the male sex organ of pteridophytes borne on the gametophyte and producing spermatozoids or antherozoid.

Apomixis: the formation of a sporophyte from a gametophyte by direct, asexual development, rather than by fertilisation of gametes.

Apophysis: swelling on which a sorus is located.

Apothecium: cup shaped fruiting body with asci.

Appendage: A secondary part attached to the main structure; an external growth that seldom has any obvious function.

Archegonium: The female sex organ of pteridophytes borne on the gametophyte and producing eggs.

Areole: an area surrounded by anastomosing veins.

B

Baculate: of spores, bearing cylindrical projections more than 1 ?m long and less in diam., and usually with an obtuse to rounded apex.

Basidiocarp: cup shaped fruiting body forming basidiospores.

Basiscopic: facing or directed towards the base of the axis on which the structure is borne.

Binomial: making use of names consisting of two words to form the scientific name (or combination) in a Latin form. For example, where the first is the name of the genus to which the species belongs, and the second is the epithet given to that species to distinguish from others in the same genus.

Binomial nomenclature: the system of nomenclature in which the scientific name of a species (and not of a taxon at any other rank) is a combination of two names, the first name being the generic name. The second name is referred to botanically as the specific epithet. Note that the two names constitute the species name, not just the second word.

Bryophyte: a botanical group including mosses and liverworts. Technically a classification of plants including three classes: hornworts (Anthocerotae), liverworts (Hepaticae) and mosses (Musci).

Bulbil: a small, usually persistent, globose, usually hairy or scaly, asexual propagule borne on a root, rhizome, or frond and capable or not capable of forming a plantlet, as in certain ferns, Asplenium, Dryopteris, Polystichum, and Woodwardia species.

C

Capsule: the part of a sporangium that contains the spores.

Catadromous: with the basiscopic pinnule and/or vein group of the pinna the first to depart from its axis.

Cleistothecium: closed fruiting body of ascomycetes.

Coenobium: colony of independent organisms united by common boundary.

Coenocytic: multinucleate without septum.

Coenosorus: a compound sorus composed of several contiguous sori fused end-to-end.

Commissure: place where a commissural vein joins parallel, otherwise free veins along the lamina margin and often underlies a continuous, marginal coenosorus.

Cone: a fruit, usually woody, ovoid to globular, including scales, bracts or bracteoles arranged around a central axis, e.g. in gymnosperms, especially conifers and casuarina.

Conidiophore: hyphal branch bearing conidia.

Costa: the major axis of a pinna; synonyms mid-rib, mid-vein.

Costate: of scales, with a central line of cells different (usually darker) from the more marginal cells, as in certain Aspleniaceae and Dryopteridaceae.

Costule: the major axis of a pinnule; synonyms mid-rib, mid-vein, rachilla.

Cryptogams: ferns, bryophytes, algae and fungi (including lichenized fungi); 'lower plants'; plants producing spores, and without stamens, ovaries or seeds, literally plants, whose sexual reproductive organs are not conspicuous cf. phanerogam.

D

Dictyostele: a siphonostele with more than one parenchymatic gap at a single level.

E

Echinate: of spores, with the outer wall (exospore or perispore) raised in a pattern of long spines.

Epiphytic: a plant which grows on another plant, but manufactures its own food.

Epispore: a particular, outer, sporopollenin layer of the spore wall that is external to, but partially attached to the exospore in some heterosporous ferns and equisetum.

Exospore: the principal sporopollenin layer of the spore wall internal to the perispore; synonym exine.

F

False indusium: an introrse, reflexed or revolute, often modified lamina margin that protects young sporangia.

False vein: an elongate series of thickened cells appearing to be a vein, but not connected to true veins and not functioning as a vein; found in the laminae of certain Aspleniaceae, Hymenophyllaceae, Marattiaceae, Pteridaceae, and Selaginellaceae.

Ferns: the pteridophytes excluding the Isoëtaceae, Lycopodiaceae, and Selaginellaceae.

Frond: the photosynthetic organ of ferns, usually consisting of a stipe and lamina; synonym leaf.

G

Gametophyte: an inconspicuous, non-vascular stage in the life cycle of a pteridophyte that bears gametangia with gametes.

Gemma: a structurally specialised, asexual propagule found on some gametophytes that detaches and forms a new gametophyte.

Gametophyte: plant that bears gametes; in ferns, usually a small but discrete plant very different from the sporophyte (which is normally considered the fern plant); in gymnosperms and angiosperms, a microscopic structure (part of the reproductive apparatus) not recognizable as a discrete plant.

Globule: male organ of Chara.

H

Haploid: of chromosomes, and relative to the phase of an alternation of generations in which the duplicated chromosome set or diploid condition is reduced; the condition when the chromosomes are not duplicated, e.g. the complement of chromosomes in the nucleus of a gamete; a single basic set of chromosomes in the nucleus of a cell. This may be expressed symbolically as n, where n = the gamete number of chromosomes.

Heterosporous: producing spores of two sizes, each of which develops gametophytes having gametangia of a single sex.

I

Indusium: a usually thin, often scalelike, epidermal membrane subtending and/or covering the sorus, that partially or fully protects the young sporangia.

Intrastelar canal: a channel occurring within a stele.

Isodromous: with the basal pinnules and/or vein groups of the pinnae strictly opposite.

Isophyllous: having fronds or microphylls of a single size and shape.

L

Laesura: the simple, elongate or triradiate, scar on the surface of pteridophyte spores; synonym suture.

Lamina: the expanded portion of a frond, usually consisting of a rachis or costâ, other axes or lateral veins, and expanded lamina tissue; synonym blade.

Leptosporangium: a thin-walled, thin-pedicelled sporangium bearing usually 64 spores (32 in apomixises of the Dryopteridaceae, 128-512 in the Osmundaceae, and 256 in the Schizaeaceae) and formed usually from a single epidermal initial cell.

Ligule: a small, tonguelike, often triangular appendage located near the microphyll base distal to the sporangium on the adaxial surface of the microsporophyll; it is persistent in Isoëtes.

Lumen: the central cavity of a cell, especially applied to the cells of clathrate rhizome scales.

Lycophytes: collectively the Isoëtaceae, Lycopodiaceae, and Selaginellaceae.

M

Massula: in Azolla, a structure derived from the contents of the microsporocarp that contains the microspores and has glochidia (minute barbed hairs) protruding from its surface.

Megagametophyte: in heterosporous pteridophytes, a female gametophyte borne within a megasporangium and bearing one or more archegonia.

Megasporangium: a sporangium bearing megaspores.

Megaspore: a large spore of the heterosporous pteridophytes Azolla, Isoëtes, Marsilea, Pilularia, Regnellidium, Salvinia, and Selaginella that produces a female gametophyte.

Megasporocarp: a sporocarp that bears megasporangia.

Megasporophyll: a fertile microphyll bearing or subtending a megasporangium.

Microgametophyte: in heterosporous pteridophytes, a male gametophyte borne within a microsporangium and bearing one or more antheridia.

Microphyll: the photosynthetic organ of the lycophytes, Equisetaceae, and Psilotaceae, always lacking a stipe, often small and generally supplied with a single vascular bundle; usually associated with a protostele or siphonostele.

Microsporangium: in heterosporous pteridophytes, a sporangium bearing microspores.

Microsporocarp: a sporocarp that bears microsporangia.

Microsporophyll: a fertile microphyll bearing or subtending a microsporangium.

Monolete: of spores, bilaterally symmetric with a linear, unbranched laesura.

Murus: of spores, an elongate, wall-like protuberance.

P

Paraphysis: a minute, unicellular or multicellular resembling a simple hair, usually elongate and sometimes glandular structure is borne on the soral receptacle, on the sporangium capsule or pedicel.

Perispore: the outermost, sporopollenin layer of the spore wall that is deposited on the exospore; synonym perine.

Phyllopodium: in ferns with articulate stipes, that portion of the stipe proximal to the articulation that remains attached to the rhizome. Phyllopodia are especially prominent and stipelike in, e.g., Elaphoglossum and Oleandra but low and more like the rhizome and knoblike in Adiantum, Davalliaceae, and many Polypodiaceae.

Pinna: a stalked or sessile, primary division of a compound lamina that is at least narrowed at the base.

Pinnule: a stalked or sessile division of a pinna that is at least narrowed at the base.

Plectostele: a vascular cylinder with the vascular tissue appearing to be parallel plates found in many Lycopodiaceae.

Primordium: a part in its most rudimentary form or stage of development.

Proliferous: forming bulbils or plantlets, often on parts that normally have another function such as roots, stems and branchlets, rachises or lamina margin.

Protostele: a simple vascular cylinder that lacks a pith of parenchyma in the center and is without parenchymatic gaps.

Pteridophytes: non-seed-bearing vascular plants; ferns and lycophytes collectively.

Pulvinus: a swollen structure at the base of a frond or at the base of pinnae, particularly common in the Marattiaceae.

R

Rachis: the principal, central axis of a pinnatifid or more compound lamina.

Receptacle: the point or region of the lamina tissue, often thickened and amply supplied by one or more veins, that produces sporangia and sometimes paraphyses and/or sporangiasters.

Rhizoid: an elongate, non-vascularised, uni- or paucicellular structure that serves to anchor the gametophyte and to absorb water and nutrients from the substrate.

Rhizome: a scaly or hairy rarely glandular or glabrous anchoring stem that bears roots and fronds.

Rhizophore: a specialised, aerial root of Selaginella that arises in the axils of stems and branches repeatedly when in contact with the substrate; it may occasionally differentiate into a stem, rather than a root.

S

Sinus membrane: an elongate translucent membrane below the sinus in a pinnule.

Siphonostele: a vascular cylinder that has a pith of parenchyma in the centre and phloem on both the outside and inside of the cylinder or in the Osmundaceae only on the outside of the cylinder.

Solenostele: a siphonostele with only one parenchymatic gap at a single level.

Soriferous: bearing sori.

Sorophore: the gelatinous, sporangium-bearing ring is produced by sporocarp as in the Marsileaceae during germination.

Sorus: a cluster of sporangia.

Spermatozoid: a motile male sex cell gamete is produced in the antheridium.

Sporangiophore: a greatly transformed, peltate sporophyll bearing a ring of sporangia face the axis of the strobilus of Equisetum.

Sporangium: the spore-producing structure of pteridophytes.

Spore: a spherical, tetrahedral, or reniform, often elaborately ornamented, reproductive cell that is produced within the sporangium and germinates to form a gametophyte.

Sporocarp: in Marsilea, the hard, short to long-pedunculate, nutlike structure containing the sporangia, apparently a highly modified leaflet; in Azolla and Salvinia, a thin, short-stalked, globose structure containing the sporangia, apparently a modified indusium.

Sporophore: the fertile portions of a hemidimorphic frond as in the Ophioglossaceae.

Sporophyll: in ferns, a frond bearing sporangia

Stele: the vascular and associated tissues of a rhizome or other type of stem.

Stipe: the structure of a frond that connects the base of the lamina to the point of its attachment to the rhizome.

Stipicel: a term is sometimes used for the stalk of a pinna or pinnule.

Stipule: in the Marattiaceae, each one of a pair of lateral, fleshy, starch-bearing, persistent, partially or entirely vascularised outgrowth of the rhizome clasp with the base of the stipe, capable of vegetatively reproducing the plants; in the Ophioglossaceae, merely the remnants of the older stipe base that originally enclosed and protected is the younger, less developed fronds.

Strobilus: in the lycophytes and Equisetaceae, a compact reproductive structure is borne at the tips of branches or axes consisting of a central axis bearing closely spaced, spirally arranged sporophylls or sporangiophores.

Synangium: a group of sporangia partially or entirely fused laterally, as in Psilotum and Marattiaceae.

T

Trilete: of spores, radially symmetric (spherical or tetrahedral) with a laesura with three radiating branches.

Trophophyll: a vegetative, nutrient-producing frond or microphyll.

Trophopod: the enlarged, persistent, basal portion of a stipe that functions as a storage organ.

V

Valve: an involucral lobe, especially in Hymenophyllum and some Dicksoniaceae and Dennstaedtiaceae; also, each half of a sporangium that is divided into halves, as in the Lycopodiaceae and Osmundaceae.

Vascular bundle: an elongate strand of conducting cells, xylem tracheids and phloem sieve cells that serve to conduct water, mineral nutrients, and photosynthetic products.